NOVEL CONCEPTS, SYSTEMS FOR SLUDGE MANAGEMENT IN EMERGENCY AND SLUM SETTINGS

Peter Matuku MAWIOO

NOVEL CONCEPTS, SYSTEMS AND TECHNOLOGY FOR SLUDGE MANAGEMENT IN EMERGENCY AND SLUM SETTINGS

DISSERTATION
Submitted in fulfilment of the requirements of
the Board for Doctorates of Delft University of Technology
and
of the Academic Board of the IHE Delft
Institute for Water Education
For
the Degree of DOCTOR
to be defended in public on
Thursday, January 16, 2020 at 12.30 hours
in Delft, the Netherlands

By

Peter Matuku MAWIOO
Master of Science in Municipal Water and Infrastructure; Specialization Sanitary Engineering,
IHE Delft Institute for Water Education, Delft, the Netherlands
Born in Kitui, Kenya

This dissertation has been approved by the
Promotor: Prof. dr. D. Brdjanovic
Copromotor: Dr. C.M. Hooijmans

Composition of the doctoral committee:

Rector Magnificus TU Delft   Chairman
Rector IHE Delft   Vice-Chairman
Prof. dr. D. Brdjanovic   IHE Delft / TU Delft, promotor
Dr. C.M. Hooijmans   IHE Delft, copromotor

Independent members:
Prof. dr. G.H. Chen   The Hong Kong University of Science and Technology,
                                                   Hong Kong
Em. Prof. dr. A.H. Maiga   2iE, Burkina Faso
Prof. dr. M. Matošić   University of Zagreb, Croatia
Prof. dr. ir. J.B. van Lier   TU Delft
Prof. dr. ir. M.K. de Kreuk   TU Delft, reserve member

Dr. H.A. Garcia of IHE Delft, the Netherlands has significantly contributed towards the supervision of this dissertation.

This research was conducted under the auspices of the Graduate School for Socio-Economic and Natural Sciences of the Environment (SENSE)

Published by:
CRC Press/Balkema
PO Box 11320, 2301 EH Leiden, The Netherlands
Pub.NL@taylorandfrancis.com
www.crcpress.com – www.taylorandfrancis.com
ISBN 978-0-367-90221-6 (Taylor & Francis Group)

To
Joan
Jayson
Jessica
Joel
and
Evelyn

Acknowledgement

I would like to express sincere gratitude to my promotor Prof. Dr. Damir Brdjanovic and my supervisors Dr. Hector A. Garcia and Prof. Christine M. Hooijmans from whom I received enormous support, encouragement, guidance and advice throughout the PhD research period. Their unprecedented efforts, critical and innovative insights and probing discussions made a great deal in shaping this research. I am grateful to the Bill and Melinda Gates Foundation for providing the funding that financed all aspects of this research. Sincere gratitude to the enitire team of Fricke und Mallah Microwave Technology GmbH (Peine, Germany) and Tehnobiro d.o.o (Maribor, Slovenia) for the the fruitful discussions and technical support during the manufacturing and testing phase of the reactor unit. Thanks Sanergy, Kenya for hosting me and providing laboratory support during my research field investigations in Nairobi, Kenya.

Much appreciation also goes to all the staff at the Department of Environmental Engineering and Water Technology (EEWT) of IHE Delft Institute for Water Education, Delft, the Netherlands. I also acknowledge the support from Prof. Thammarat Koottatep and the environmental laboratory staff of the Asian Institute of Technology (AIT), Thailand, and Prof. Marjana Simonič and the laboratory staff of Maribor University, Slovenia for support in sample analysis. I gratefully acknowledge the MSc participants that made enormous contributions to this study, particularly Audax Rweyemamu, and Mary Barrios Hernandez.

I thank my PhD colleagues for the discussions and time (including the fun) that made life easier during the study, in particular Joy Riungu, Fiona Zakaria, Yuli Ekowati, Laurens Welles, Javier Sanchez, Sondos Saad, Nikola Stanic, Jeremiah Kiptala, Micah Mukolwe, Frank Masese, Nirajan Dhakal, Abdulai Salifu, Chol Abel, Nadejda Andreev, and Josip Čurko.

Lastly, I would most thank my family for the incredible support and allowing me to be away during the study, particularly my wife Ms. Evelyn Nduku, my daughters Jessica Mutheu and Joan Kasili and my sons Joel Muthui and Jayson Mumo - Junior. I am grateful to my father Mzee Philip Mawioo and my late mother Ms. Linah Kasili Mawioo for the sacrifices they made for my earlier education that has culminated in this thesis.

Peter Matuku Mawioo
Delft, December 10, 2019

Summary

Management of sludge is one of the most pressing issues in sanitation provision. The situation is especially complex when large quantities of fresh sludge containing various contaminants are generated in onsite sanitation systems in urban slums, emergency settlements and wastewater treatment facilities that require proper disposal of the sludge. The application of fast and efficient sludge management methods is important under these conditions. This study focuses on addressing the existing challenges and gaps in sludge management, particularly the management of faecal sludge that is generated in the densely populated areas, through innovative concepts and technological development. To assess the current status of decentralized management of faecal sludge, a review of the existent (emergency) sanitation practices and technologies was conducted. In the study, the gaps and opportunities in technological developments for sanitation management in complex situations was identified. The need for an innovative sludge management system led to the development of the "emergency sanitation operation system, eSOS". This concept proposed and demonstrated the application of modern innovative sanitation solutions and existing information technologies for sludge management. In addition, as a component of the eSOS concept, a sludge treatment system based on microwave irradiation technology, which forms the core of this research, was developed and tested. The microwave technology study was carried out in two stages. The first stage involved preliminary and validation tests at laboratory scale using a domestic microwave unit to assess the applicability of the microwave technology for sludge treatment. Two sludge types, namely blackwater sludge, extracted from highly concentrated raw blackwater stream, and faecal sludge, obtained from urine diverting dry toilets, were tested. The results demonstrated the capability of the microwave technology to rapidly and efficiently reduce the sludge volume by over 70% and decrease the concentration of bacterial pathogenic indicator $E.$ $coli$ and $Ascaris$ $lumbricoides$ eggs to below the analytical detection levels.

Basing on these results, a pilot-scale microwave reactor unit was designed, produced and evaluated using waste activated sludge, faecal sludge, and septic sludge, which formed the second stage of the study. The results demonstrated that microwave treatment was successful to achieve a complete bacterial inactivation like in the laboratory tests (i.e. $E.$ $coli$, coliforms, $staphylococcus$ $aureus$, and $enterococcus$ $faecalis$) and a sludge weight/volume reduction above 60%. Furthermore, the dried sludge and condensate had a high energy (≥ 16 MJ/kg) and nutrient contents (solids; TN ≥ 28 mg/g TS and TP ≥ 15 mg/g TS; condensate TN ≥ 49 mg/L TS and TP ≥ 0.2 mg/L), having the potential to be used as biofuel, soil conditioner, fertilizer, etc.

Overall, in this study the existence of a wide range of regular onsite and offsite sanitation options was revealed that have the potential to be applied for sludge management in the emergencies. Situations with more or less similar characteristics to emergencies such as urban slums can also benefit from these technologies.

In addition, the shortfalls experienced in the many current emergency sanitation responses were associated with the often used conventional fragmented approach that does not capture the entire sanitation chain, but rather looks at the individual components separately with emphasis on the containment facilities. An innovative emergency Sanitation Operation System (eSOS) concept was thus introduced in this study that uses and promotes a systems approach integrating all components of an emergency sanitation chain.

Furthermore, the study demonstrated that a microwave technology based reactor can be applied for the rapid treatment of sludge in the areas where large volumes of sludge are generated such as slums and emergency settlements.

Samenvatting

Het beheer van slib is een van de meest dringende kwesties in sanitatie voorzieningen. De situatie is des te meer complex wanneer grote hoeveelheden vers slib met verschillende verontreinigingen worden gegenereerd in afvalwaterzuiveringsinstallaties en sanitatie systemen in sloppenwijken en vluchtelingenkampen, waarvan het slib naar behoren moet worden behandeld.

De toepassing van snelle en efficiënte methoden voor slibbeheer is belangrijk onder deze omstandigheden. Deze studie richt zich op het aanpakken van de bestaande uitdagingen en hiaten in het beheer van fecaal slib in deze dichtbevolkte gebieden door middel van innovatieve concepten en technologische ontwikkeling. Om de huidige status van gedecentraliseerd beheer van fecaal slib te beoordelen, werd een evaluatie van de bestaande (nood) sanitatiepraktijken en technologieën uitgevoerd. In de studie werden de hiaten en kansen voor technologische ontwikkelingen voor sanitatiebeheer in complexe situaties geïdentificeerd. De behoefte aan een innovatief managementsysteem voor slib leidde tot de ontwikkeling van het "noodsysteem voor sanitaire voorzieningen, eSOS". Dit concept laat de toepassing van moderne innovatieve sanitatieoplossingen en bestaande informatietechnologieën voor slibbeheer zien. Daarnaast werd als onderdeel van het eSOS-concept een slibbehandelingssysteem op basis van microgolfbestraling, dat de kern vormt van dit onderzoek, ontwikkeld en getest. Het onderzoek naar microgolftechnologie werd in twee fasen uitgevoerd. De eerste fase omvatte voorlopige en validatieproeven op laboratoriumschaal met behulp van een huishoudelijke magnetron om de toepasbaarheid van de microgolftechnologie voor slibbehandeling te beoordelen. Twee slibtypes, namelijk slib geëxtraheerd uit een geconcentreerde zwart-water stroom, en fecaal slib, verkregen uit urine gescheiden, droge toiletten, werden getest. De resultaten lieten het vermogen van de microgolftechnologie zien om het slibvolume snel en efficiënt met meer dan 70% te verminderen en de concentratie van bacteriële pathogene indicator *E. coli* en *Ascaris lumbricoides* eieren tot onder de analytische detectieniveaus te verlagen.

Op basis van deze voorlopige resultaten werd een microgolfreactor op proefschaal ontworpen en geproduceerd, en de werking geëvalueerd met behulp van actief slib, fecaal slib en septisch slib, dat de tweede fase van het onderzoek vormde. Op dezelfde manier hebben de resultaten aangetoond dat microgolfbehandeling succesvol was om een volledige bacteriële inactiviteit te bereiken zoals in de laboratoriumtests (dat wil zeggen *E. coli*, coliformen, *staphylococcus aureus en enterococcus faecalis*) en een slibgewicht / volumeafname van meer dan 60%. Bovendien had het gedroogde slib en het condensaat een hoge energiewaarde (≥ 16 MJ / kg) en nutriënten concentratie (vaste stof; TN ≥ 28 mg / g TS en TP ≥ 15 mg / g TS; condensaat TN ≥ 49 mg / L TS en TP $\geq 0,2$ mg / L), met mogelijke toepassing als biobrandstof, bodemverbeteraar, kunstmest, etc.

In dit onderzoek werd het bestaan onthuld van een breed scala van reguliere on- en off-site beschikbare sanitatie opties die in potentie kunnen worden toegepast voor slibbeheer in noodsituaties. Situaties met min of meer vergelijkbare kenmerken als van noodsituaties, zoals sloppenwijken in de stad, kunnen ook van deze technologieën profiteren.

Bovendien bleek dat de sanitatie tekortkomingen ondervonden in de vele huidige noodsituaties geassocieerd zijn met de vaak gebruikte conventionele, gefragmenteerde aanpak die niet de hele sanitatieketen beschouwd, maar de afzonderlijke onderdelen. In dit onderzoek werd een innovatief concept, eSOS (Sanitation Operation System), voor noodgevallen geïntroduceerd, dat een systeembenadering gebruikt en promoot, waarbij alle componenten van noodhulp sanitatie worden geïntegreerd. Het liet zien dat door het promoten van een systeembenadering, het eSOS-concept de vooringenomenheid kan aanpakken bij het ontwikkelen van de componenten van de sanitatieketen.

Tenslotte toonde de studie aan dat een op microgolftechnologie gebaseerde reactor kan worden toegepast voor de snelle behandeling van slib in de gebieden waar grote volumes slib worden gegenereerd, zoals sloppenwijken en vluchtelingen kampen.

Table of contents

List of figures

List of tables

Chapter 1
Introduction

1.1 General introduction

There has been concerted efforts at both international and local levels to increase access to improved sanitation facilities that ensure hygienic separation of human excreta from human contact. However, according to an update by the WHO/UNICEF Joint Monitoring Programme for Water Supply, Sanitation and Hygiene (JMP) for the year 2017, approximately 2.3 billion people around the world still lacked access to basic sanitation services, while some 892 million practiced open defecation (WHO/UNICEF, 2017). The consequences of poor sanitation are widely acknowledged. For instance, it has been reported that over 80 percent of all diseases in the developing world result from poor sanitation and contaminated drinking water resources (Afolabi and Sohail, 2017). It is also reported that sanitation related diarrhoeal infections kill around 700,000 children every year (UN Inter-Agency Group for Child Mortality Estimation, 2012). Various studies have demonstrated that improved sanitation can reduce the diarrhoeal infections by up to 60 percent (Walker et al.; Esrey et al., 1991; Norman et al., 2010; Munamati et al., 2016). One of the main areas of concern regarding sanitation is the management of human excreta, especially in the locations where large quantities are generated that require frequent emptying. For instance, the onsite sanitation technologies, particularly portable toilets, pit latrines, septic tanks, etc., which are commonly applied in the densely populated areas (e.g. emergency settlements and slums) (Harvey, 2007; Katukiza et al., 2012; Brdjanovic et al., 2015) are normally intensively used and frequently emptied resulting in large quantities of highly contaminated fresh faecal and septic sludge. Factors such as time and land space constraints (see Figure 1-1), etc., are also frequently encountered in those isolated situations, which further complicates an already challenging task of sludge management.

Figure 1-1. Kutupalong camp at Cox's Bazar, Bangladesh, March 2019 (Photo: C.M. Hooijmans)

In most cases, the impacts of poor sanitation are higher in the emergency settlements owing to the usual destruction of the existent water and sanitation infrastructure in the aftermath of a disaster event and the consequent sudden agglomeration of the affected people in the emergency camps.

1.2 Sanitation in emergency situations

Generally, inadequate sanitation provision has a wide range of consequences that are well acknowledged, but the implications are more aggravated in isolated conditions such as the emergencies and the poorly developed urban settings with high dense habitation (informal settlements), which can also be considered as permanent emergencies. The importance of sanitation in disaster conditions is underscored in the Sphere Project: humanitarian charter and minimum standards in humanitarian response (The Sphere Association, 2018), which identifies water and sanitation as critical determinants for survival in the early stages of an emergency. People affected by a disaster are vulnerable to illness and diseases which are largely related to inadequate sanitation and water supplies as well as poor hygiene. The most significant of these diseases are diarrheal and infectious diseases that are mainly transmitted via the faecal-oral or skin penetration routes as illustrated in Figure 1-2 (Harvey et al., 2002).

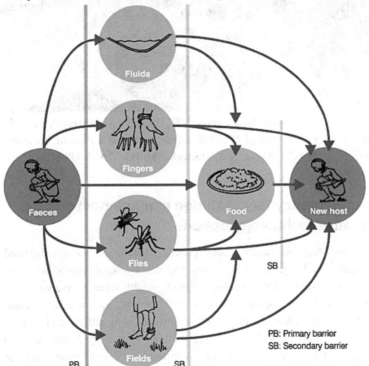

Figure 1-2. Disease transmission from faeces (Harvey et al., 2002, adapted from Kawata, 1978)

Therefore, adequate intervention measures are necessary for emergency sanitation to curb or reduce the transmission or spread of the water and sanitation related diseases (Harvey et al., 2002). Generally, emergency sanitation comprises several components including faecal sludge or excreta disposal, vector control, solid waste disposal and drainage (The Sphere Association, 2018). However, most sanitation related disease outbreaks and spread can be linked to the excreta disposal, particularly the faecal sludge. Excreta related infections in emergencies can be linked to both the lack of better approaches and appropriate (technology) options for faecal sludge containment, emptying and transportation, and the treatment of the faecal sludge generated from the commonly applied onsite sanitation facilities such as toilets and other options (e.g. peepoo and wagbags, etc.). The problems around emergency sanitation as discussed above are widely acknowledged by the humanitarian actors as demonstrated in their continuous efforts to develop a range of faecal sludge management solutions.

This is illustrated by the recent effort in Cox's Bazar, Bangladesh. See Figure 1-3 for two different on-site systems based on chemical (lime) treatment and biological (aerobic) treatment. Lime treatment is the most applied decentralised technology. Other technologies that are tested in Cox's Bazar are e.g. biogas tanks and anaerobic filters.

Figure 1-3. Lime treatment of the British Red Cross (left, Rohinya volunteers), aerobic treatment of the International Federation of the Red Cross and Red Crescent Societies (right). Kutupalong camp at Cox's Bazar, Bangladesh, March 2019 (Photo: C.M. Hooijmans)

1.3 Challenges in emergency faecal sludge management and the need for alternative technology options

The management of large quantities of faecal sludge generated from the intensively used sanitation facilities is a great challenge in emergencies and other similar conditions. The handling and disposal processes of the large sludge volumes might incur excessive costs. Besides, fresh faecal sludge contains high amounts of pathogens (e.g. bacteria, viruses, protozoa, and helminths (WHO, 2001; Jimenez et al., 2006; Fidjeland et al., 2013)) and organic matter (Mawioo et al., 2016a; Mawioo et al., 2016b), and if poorly managed may lead to contamination of ground and water sources. These may in turn act as transmission points of infectious disease and pathogens to people who come in contact with them. Furthermore, faecal sludge may provide breeding sites for vectors like flies and mosquitoes that can transmit disease. Vermin and domestic animals may also be attracted by the faecal sludge which in turn may increase the potential for diseases (Harvey et al., 2002; John Hopkins and IFRC, 2008). Aesthetically, improper disposal of faeces can create unpleasant environmental conditions in terms of odor and sight. These concerns require to undertake fast yet appropriate actions to safeguard the affected people whose public health is at risk in the poor disaster conditions.

Unfortunately, the adoption of the appropriate solutions is not always possible and in some emergencies, the outbreaks of the excreta-related epidemics have been linked with inadequate sanitation provision. Some recent incidences include the rapid spread of cholera after the Haiti earthquake in 2010 (Tappero and Tauxe, 2011) and the most recent floods following the cyclone Idai in Mozambique in 2019 (Beaumont, 2019, March 27), the outbreaks of diarrheal epidemics after the Pakistani earthquake in 2005, the tsunami in Indonesia in 2004, the floods in Bangladesh in 2004, and the floods in Mozambique in 2000 (Watson et al., 2007). Furthermore, John Hopkins and IFRC (2008) identified diarrheal infections among the common causes of mortality in emergency situations. They accounted for approximately 25 percent of the deaths reported in Kohistan district, Afghanistan between November 2000 and April 2001. Moreover, over 85 percent of the deaths among the Rwandan refugees in Goma, Zaire, following the Rwandan genocide in 1994 (Harvey, 2007), and 41 percent of the under-five child mortality in an eastern Ethiopia refugee camp (Davis and Lambert, 2002) were associated with diarrheal diseases caused by poor sanitation practices.

A range of current conventional technology options for faecal sludge containment, emptying and transportation is applicable in emergencies. However, the treatment component is particularly challenging since rapid processes are required that can match the large quantities of faecal sludge that often are emptied from the containment facilities when hardly any organic degradation and pathogen die-off process has occurred. A number of conventional treatment options for faecal sludge are available, including drying (in sludge drying beds), composting, co-digestion with solid waste (producing biogas), and co-treatment in wastewater treatment plants (Ingallinella et al., 2002; Robert et al., 2018). Although they were successfully tested and applied in the regular sanitation context, these technologies have various limitations to their application in the emergency contexts. For instance, the composting technology produces a hygienically safe product rich in humus carbon, fibrous material, nitrogen, phosphorus and potassium. However, it has limitations including large space requirements, long treatment duration, and environmental pollution and public health concerns in low-lying areas in case of flooding (Katukiza et al., 2012). Co-digestion with solid waste offers the benefit of increased yield of biogas and use of the end product as a fertilizer, but has the limitation that it involves relatively slow conversion processes and requires post treatment stage for further pathogens destruction (Katukiza et al., 2012). The co-treatment in conventional wastewater treatment plants is a possible option for faecal sludge treatment but the likely organic, nutrients and solids overload may be a drawback, especially if it was not planned in the design (Lopez-Vazquez et al., 2014). Furthermore, wastewater treatment plants might be destroyed during the disaster event or simply do not exist in the vicinity of an emergency settlement. The major limitations of the conventional faecal sludge treatment technologies discussed above are largely related to their relatively slow treatment processes and large land space requirements making them less feasible to apply in the critical stages of emergency scenarios, which are often characterized by high generation rates of faecal sludge, limited land space, and time constraints. Currently lime treatment, a non-biological treatment process is largely applied in Cox's Bazar, Bangladesh. However, although bacterial inactivation is taking place, helminths seem to survive the treatment, and for drying of the sludge large land space is required.

Generally, the challenges and the resulting public health incidences observed in the past emergencies demonstrate the need to further enhance the provision of emergency sanitation. On this basis, the development of innovative technologies, especially for the emergency faecal sludge treatment, is necessary. In recent years, there have been remarkable efforts to expand the onsite toilet options, including the emergency toilet options, in which a number of prototypes have been developed. However, parallel efforts to develop technology options to rapidly and effectively treat faecal sludge generated from those toilets still have to be demonstrated in practice. Recently, a faecal sludge treatment technology known as LaDePa (Latrine Dehydration and Pasteurization) machine has been developed and successfully evaluated under the regular development context (Septien et al., 2018) but, not tested under emergency conditions.

To ensure optimal benefits, the development of the treatment technologies for the faecal sludge management in disaster situations should be matched with the three globally adopted emergency phases namely; (i) phase one, of duration up to two weeks, (ii) phase two, of up to six months duration, and (iii) phase three, lasting in excess of six months. This is because the different phases have different characteristics that would need relatively different approaches, which is the reason the design of emergency sanitation programmes are staged to correspond with the respective phases. The first phase may be regarded as the most critical concerning sanitation provision due to the related chaotic environment resulting from the sudden agglomeration of the affected people in unplanned settlements. At this stage, sanitation provision is aimed at containing spread of sanitation related diseases and ensure safer environment to protect human life and health. Due to the urgency for the provision of sanitation services at this stage, the corresponding potential technology options should be designed to meet several requirements, particularly compactness, ease of deployment, rapid and highly efficient treatment processes, and modular for easy installation, handling and transportation. The design of the faecal sludge treatment technology should envisage aspects such as pathogen inactivation, volume reduction and reuse. However, sanitization (i.e. inactivation of pathogenic organisms) of the faecal sludge should be given first priority to ensure public health safety, after which volume reduction and reuse can be considered to minimize handling and disposal costs as well as promote resource recovery. In conclusion, emergency sanitation sector is still awaiting for the **rapidly employable** technology which can **quickly** and **efficiently sanitize** faecal sludge. This was the main driver behind this PhD thesis research that focuses on a technology based on the microwave irradiation which has potential to achieve the requirements discussed above (Mawioo et al., 2016a).

The microwave based heating technologies rely on the microwave energy, which is a part of the electromagnetic spectrum with wavelengths (λ) ranging from 1 mm to 1 m and frequencies between 300 MHz (λ =1m) and 300 GHz (λ =1mm) (Haque, 1999; Tang et al., 2010; Remya and Lin, 2011). Heating of a material by microwaves results from the rotation of dipolar species and/or polarization of ionic species due to their interaction with the electromagnetic field (Haque, 1999). The molecular rotation and migration of ionic species causes friction, collisions, and disruption of hydrogen bonds within water; all of which result in the generation of heat (Venkatesh and Raghavan, 2004). Faecal sludge and other types of sludge contain high amount of dipolar molecules such as water and organic complexes, which makes them good candidates for the microwave dielectric heating. The technology can thus be investigated to determine its potential to be applied for the treatment of faecal sludge in emergencies and other highly populated conditions.

6

It is also important to point out that technology (i.e. hardware) alone cannot address and solve all challenges around emergency sanitation. Often, sometimes complex relations between components in sanitation service provision chain are not sufficiently defined and understood, which leads to suboptimal solutions. Therefore, development of a decision support framework (ideally embedded in an appropriate software) is required to support the choices made and improve operation and maintenance of the sanitation systems because it is equally important to ensure proper use and functioning of technologies. For example, application of a dedicated framework for the selection of (set of) the most appropriate emergency technology option(s) based on a multi-criteria analysis is necessary to ensure that the best options are chosen. In addition, to ensure optimal functioning of the sanitation system the major components of the sanitation chain including containment, emptying, transport, treatment and safe disposal or reuse need to be considered during the planning stage. Therefore, an emergency sanitation management concept is required that promotes a systems approach in which all parts of the sanitation chain are considered holistically and viewed as integral parts of the whole system. In such a concept, novel microwave based technology can become an important integral part of the emergency sanitation system addressing the treatment component of it.

1.4 Aim and scope of the thesis

The PhD thesis aims at addressing the existing challenges and gaps in sludge management through innovative concepts and technological development. The first objective of the research was to find out which technological options are most suitable for emergency sanitation and to collect the evidence for better understanding of the decision making process regarding the technological choices made in the provision of emergency sanitation. The second objective, inspired by the outcome of the emergency conference held in Delft in 2014, was to rethink the emergency sanitation service delivery chain and to find out how it can be improved. The third, and final objective was to investigate whether the microwave-based technology for sterilizing and drying of faecal sludge can be considered feasible for application in emergency situations. Consequently, these objectives have determined the scope of this thesis.

Although the issue of sludge management is evident in many locations, the current research focuses mainly on the concepts and technological development relevant for the management of faecal sludge from onsite sanitation systems located in the densely populated areas such as emergency settlements and slums. However, the proposed concept and technology can also be applied for sludge handling in the centralized systems such as the wastewater treatment facilities. As an initial step in this research, it was necessary to assess the current status of emergency sanitation, hence a review of the existent emergency sanitation practices and technologies has been conducted. Then, a faecal sludge management concept dubbed "emergency Sanitation Operation System, eSOS", which is based on modern innovative sanitation solutions and existing information technologies, has been introduced. In addition, as a component of the eSOS concept, a faecal sludge treatment system based on the microwave irradiation technology, which forms the core of this research, has been developed and tested. The microwave technology study was carried out in two stages. The first stage involved preliminary and validation tests at laboratory scale to assess the applicability of the microwave technology for sludge treatment. A domestic microwave was used at this stage in which two tests were carried out: first using blackwater sludge extracted from highly concentrated raw blackwater from a demonstration plant in Sneek, the Netherlands and then using faecal sludge obtained from urine diverting dry toilets located in the slums of Nairobi, Kenya. In each of the tests, the capability of the microwave irradiation to inactivate pathogenic organisms and reduce sludge volume and organic matter was assessed. The information derived from the first stage informed the second stage in which a pilot-scale microwave reactor unit was designed, produced and evaluated using activated sludge, faecal sludge, and septic sludge. Similarly, pathogen inactivation, sludge volume and organic matter reduction capabilities were assessed in addition to determining the value addition of the drying end products such as the calorific and nutrition value of the dried sludge and the resulting condensate.

1.5 Research hypotheses

In this study a review of various sanitation technology options has been conducted to understand their potential application in emergency situations. An understanding of the technology selection processes used in the past emergencies was sought by conducting an overview of the documented cases and subsequently a framework for technology selection has been proposed. In addition, a concept for faecal sludge management based on a systems approach that integrates all components of emergency sanitation chain has been introduced. As part of the concept, a microwave based reactor has also been developed and its applicability in emergency sanitation and similar situations investigated. The various aspects outlined above were investigated to verify the following hypotheses:

- A large number of the regular technology options that are usually applied for sanitation provision in the development contexts have potential to be applied in emergency contexts as well.
- Current deficiencies in the selection of emergency sanitation technology options are a result of the lack of structured procedure and can be addressed by developing a dedicated emergency sanitation technology selection framework.
- An integrated systems approach in emergency sanitation provision that makes advantage of modern means and tools can improve operation and maintenance of the entire system.
- Treatment of faecal sludge by the use of microwave irradiation can inactivate pathogens present in the sludge, reduce sludge volume and its weight and hence lead to a reduction of public health risks, safe handling of treated sludge, and lower transportation cost to final disposal site.

It is expected that the exploration and adoption of the above mentioned aspects can increase the effectiveness and efficiency in the emergency sanitation response and largely contribute in providing a solution to the many issues facing sanitation provision in the emergencies and other similar situations.

1.6 Objectives

This PhD thesis focusses on understanding the underlying problems and developing, testing and scaling up solutions to the issues relating to faecal sludge management in densely populated areas such as emergency settlements and slums. The specific objectives of the research are as outlined below.

i. To conduct a comprehensive review of the faecal sludge management practices in emergency situations.

ii. To develop an innovative faecal sludge management concept based on systems approach for application in emergency situations.

iii. To conduct a preliminary study to investigate at laboratory scale the potential of a microwave based technology for treating faecal sludge.

iv. To evaluate the potential of a microwave based technology for emergency or slum sanitation applications by treating fresh faecal sludge obtained from toilets located in such locations.

v. To design and develop a pilot-scale microwave based reactor unit for faecal sludge treatment in emergency settlements and other highly populated areas.

vi. To evaluate the performance of the pilot-scale microwave based reactor unit using various kinds of sludge and determine the value addition of the process end products.

1.7 Outline

This thesis comprises eight main chapters. The current chapter (Chapter 1) provides background information on sludge management in areas of high sludge generation and introduces the broad problem which justifies the relevance of this study. This chapter also presents research hypotheses, goal, and objectives of the study. Chapter 2 presents a state of art review focussing on technology options for faecal sludge management in emergency situations and the future perspectives, while in Chapter 3 an innovative concept for the management of the faecal sludge in emergency situations is provided. Chapter 4 proposes microwave technology as a potential option for faecal sludge treatment in highly populated areas, especially the slums and emergency settlements. In Chapter 5 and Chapter 6, preliminary evaluations for the applicability of the microwave technology as an option for sludge treatment are conducted at laboratory scale. The results of those preliminary investigations form the basis for the design and development of a pilot-scale microwave reactor unit whose applicability is evaluated using various sludge types as presented in Chapter 7.

The main conclusions and implications of this research as well as directions for future research and development of the microwave technology are discussed in Chapter 8.

References

Afolabi, O.O.D., Sohail, M., 2017. Microwaving human faecal sludge as a viable sanitation technology option for treatment and value recovery – A critical review. Journal of Environmental Management 187, 401-415.

Beaumont, P., 2019, March 27. Cyclone Idai crisis deepens as first cases of cholera confirmed in Mozambique., The Guardian, Retrieved from https://www.theguardian.com).

Brdjanovic, D., Zakaria, F., Mawioo, P.M., Garcia, H.A., Hooijmans, C.M., Ćurko, J., Thye, Y.P., Setiadi, T., 2015. eSOS® – emergency Sanitation Operation System. Journal of Water Sanitation and Hygiene for Development 5, 156-164.

Davis, J., Lambert, R., 2002. Engineering in emergencies: *A practical guide for relief workers*, 2nd ed. ITDG Publishing 103–105 Southampton Row, London WC1B 4HL, UK.

Esrey, S.A., Potash, J.B., Roberts, L., Shiff, C., 1991. Effects of improved water supply and sanitation on ascariasis, diarrhoea, dracunculiasis, hookworm infection, schistosomiasis, and trachoma. Bulletin of the World Health Organization 69, 609-621.

Fidjeland, J., Magri, M.E., Jönsson, H., Albihn, A., Vinnerås, B., 2013. The potential for self-sanitisation of faecal sludge by intrinsic ammonia. Water Res. 47, 6014-6023.

Haque, K.E., 1999. Microwave energy for mineral treatment processes—a brief review. Int. J. Miner. Process. 57, 1-24.

Harvey, P.A., 2007. Excreta Disposal in Emergencies: a Field Manual. Water, Engineering and Development Centre (WEDC), Loughborough University, Leicestershire, UK.

Harvey, P.A., Baghri, S., Reed, R.A., 2002. Emergency Sanitation: Assessment and programme design. WEDC, Loughborough University, UK, Leicestershire, UK., p. 349.

Ingallinella, A.M., Sanguinetti, G., Koottatep, T., Montanger, A., Strauss, M., 2002. The challenge of faecal sludge management in urban areas - strategies, regulations and treatment options. Water Sci. Technol. 46, 285-294.

Jimenez, B., Austin, A., Cloete, E., Phasha, C., 2006. Using Ecosan sludge for crop production. Water Sci. Technol. 54, 169-177.

John Hopkins and IFRC, 2008. Water, sanitation and hygiene in emergencies in: John Hopkins and IFRC (Ed.), Public Health Guide for Emergencies, 2nd ed. John Hopkins and IFRC, pp. pp. 372-434.

Katukiza, A.Y., Ronteltap, M., Niwagaba, C.B., Foppen, J.W.A., Kansiime, F., Lens, P.N.L., 2012. Sustainable sanitation technology options for urban slums. Biotechnol. Adv. 30, 964–978.

Lopez-Vazquez, C.M., Dangol, B., Hooijmans, C.M., Brdjanovic, D., 2014. Co-treatment of Faecal Sludge in Municipal Wastewater Treatment Plants, in: Strande, L., Ronteltap, M., Brdjanovic, D. (Eds.), Faecal Sludge Management - Systems Approach Implementation and Operation. . IWA Publishing, London, UK, pp. 177-198.

Mawioo, P.M., Hooijmans, C.M., Garcia, H.A., Brdjanovic, D., 2016a. Microwave treatment of faecal sludge from intensively used toilets in the slums of Nairobi, Kenya. J. Environ. Manage. 184, Part 3, 575-584.

Mawioo, P.M., Rweyemamu, A., Garcia, H.A., Hooijmans, C.M., Brdjanovic, D., 2016b. Evaluation of a microwave based reactor for the treatment of blackwater sludge. Sci. Total Environ. 548–549, 72-81.

Munamati, M., Nhapi, I., Misi, S., 2016. Exploring the determinants of sanitation success in Sub-Saharan Africa. Water Res. 103, 435-443.

Norman, G., Pedley, S., Takkouche, B., 2010. Effects of sewerage on diarrhoea and enteric infections: a systematic review and meta-analysis. The Lancet Infectious Diseases 10, 536-544.

Remya, N., Lin, J.-G., 2011. Current status of microwave application in wastewater treatment—A review. Chem. Eng. J. 166, 797-813.

Robert, G., Amy, J., Samuel, R., Philippe, R., 2018. Compendium of Sanitation Technologies in Emergencies.

Septien, S., Singh, A., Mirara, S.W., Teba, L., Velkushanova, K., Buckley, C.A., 2018. 'LaDePa' process for the drying and pasteurization of faecal sludge from VIP latrines using infrared radiation. South African Journal of Chemical Engineering 25, 147-158.

Tang, B., Yu, L., Huang, S., Luo, J., Zhuo, Y., 2010. Energy efficiency of pre-treating excess sewage sludge with microwave irradiation. Bioresour. Technol. 101, 5092-5097.

Tappero, J.W., Tauxe, R.V., 2011. Lessons Learned during Public Health Response to Cholera Epidemic in Haiti and the Dominican Republic. Emerging Inefectious Diseases 17, 2087–2093.

The Sphere Association, 2018. The Sphere Handbook: Humanitarian Charter and Minimum Standards in Humanitarian Response, 4rd ed. Practical Action Publishing, Rugby, United Kingdom, p. 406.

UN Inter-Agency Group for Child Mortality Estimation, 2012. Levels and trends in child mortality report. UNICEF, New York

Venkatesh, M.S., Raghavan, G.S.V., 2004. An Overview of Microwave Processing and Dielectric Properties of Agri-food Materials. Biosystems Engineering 88, 1-18.

Walker, C.L.F., Rudan, I., Liu, L., Nair, H., Theodoratou, E., Bhutta, Z.A., O'Brien, K.L., Campbell, H., Black, R.E., Global burden of childhood pneumonia and diarrhoea. The Lancet 381, 1405-1416.

Watson, J.T., Gayer, M., Connolly, M.A., 2007. Epidemics after Natural Disasters. Emerging Infectious Diseases 13, 1-5.

WHO, 2001. Excreta-related infections and the role of sanitation in the control of transmission, in: Lorna, F., Jamie, B. (Eds.), Water Quality: Guidelines, Standards & Health: Assessment of Risk and Risk Management for Water-Related Infectious Disease. IWA Publishing, London, UK., p. 424.

WHO/UNICEF, 2017. Progress on Sanitation and Drinking Water e 2015 Update and SDG Baseline. WHO, Geneva.

Chapter 2
Emergency sanitation: A review of potential technologies and selection criteria

This chapter is based on:

Mawioo, P.M., Garcia, H.A., Hooijmans, C.M., Brdjanovic, D., 2018. Faecal sludge management in emergencies: A review of technology options and future perspectives. Submitted to Journal of Water Sanitation and Hygiene for Development.

Mawioo, P.M., Igbinosa E., Garcia H., Hooijmans C.M., Brdjanovic D., 2016. Emergency sanitation: A review of potential technologies and selection criteria - *In Proceedings: 3rd IWA Development Congress and Exhibition,* Nairobi, Kenya, 14-17 October 2013.

Abstract

The provision of sanitation services is one of the immediate requirements in the aftermath of a disaster. Lack of proper sanitation in an emergency situation may cause more deaths than directly caused by the actual disaster. A wide range of potential emergency sanitation technologies exists, but among others, the lack of a suitable decision support system might lead to the choice of inappropriate solutions. This article aims at providing a comprehensive review of the existent and potential emergency sanitation technologies, and providing a new decision support system for sanitation technology selection in emergency relief. Onsite and offsite sanitation technologies suitable for use in emergency relief are reviewed. A description of past sanitation interventions in emergency situations is presented to reveal sanitation solutions supplied in recent practices. This information is subsequently used to develop and present selection criteria to facilitate a multi criteria analysis in the emergency sanitation sector. It is based on an inclusive approach for the selection of the most suitable emergency sanitation technology using identified location (area-specific) criteria for technology screening and technology-specific evaluation criteria. This article reveals that only a few sanitation technologies with potential for application in emergency situations were used in the recent practices, which were generally not the most suitable alternatives. The systematic approach aids at deciding the most suitable technology in the emergency context.

2.1 Introduction

Natural and anthropogenic hazards such as earthquakes, storms, floods, wars, volcano eruptions, among others, can cause a disaster threatening the lives of people. Disasters may in turn result in emergencies if their disruptions stretch beyond the coping capacity of the society raising the need for external assistance (Davis and Lambert, 2002; WHO, 2002). One probable consequence of a disaster is the destruction of the existing sanitation facilities resulting to limited access to sanitation services and consequently the rise in public health risks. For instance, the rapid spread of a cholera epidemic after the Haiti earthquake in 2010 which claimed about 500,000 lives was associated with inadequate sanitation provision (Tappero and Tauxe, 2011). Furthermore, sanitation related outbreaks of diarrheal diseases were reported after the earthquake in Pakistan in 2005, the tsunami in Indonesia in 2004, the floods in Bangladesh in 2004, and the floods in Mozambique in 2000 (Watson et al., 2007). These incidences identify sanitation provision as one of the first priorities to which fast response is required to avert possible epidemics in emergencies. A sanitation technology review conducted here reveals existence of a wide range of technologies with potential for use in emergency response. However, the planners in the humanitarian relief often tend to rely on standard remedies and replicate similar solutions even for clearly different scenarios. This is observed when reviewing the sanitation interventions in the past major emergencies in which the majority of the technologies used were based on the traditional pit latrine and/or its variations. The reliance on standard remedies is influenced by several (non-scientific based) factors such as the ease of access to certain technologies, the tradition in practice, and the existing relations between suppliers and relief agencies. It is further exacerbated in the selection process as there is no thorough emergency-specific framework in place yet to guide the relief planners and decision makers while selecting the sanitation technologies. These deficiencies may lead to implementation of the less adequate solutions (Fenner et al., 2007), thus undermining the response efforts.

The use of a dedicated multi criteria analysis (MCA) can reduce the influence of deficiencies resulting from the application of standard remedies and intuition of the relief planners by introducing a scientific based approach in the technology selection process. Several general sanitation decision support systems (DSS) have been developed in the past (Finney, 1998; Loetscher and Keller, 2002; Zurbrügg and Tilley, 2007; Palaniappan et al., 2008; van Buuren, 2010); however, these systems primarily address sanitation in a development context. Some criteria used in these systems (such as the willingness to pay and institutional arrangements) may not be fully relevant or applicable to emergency contexts. (Fenner et al., 2007) developed a preliminary process selection tool for sanitation in disaster relief. In this tool faecal sludge and wastewater management are evaluated separately across the given range of sanitation options. Moreover, the site conditions are only considered when selecting the faecal sludge-disposal options while the selection of the wastewater-disposal options is fully based on technology characteristics, not considering the site conditions. Furthermore, the resulting competing options for faecal sludge disposal are not evaluated to the extent that would lead to the selection of the most appropriate solution. An accurate multi criteria framework would require both the consideration of the technological characteristics of the available sanitation options, as well as the local (area-specific) conditions of the disaster area.

The paper aims at providing a comprehensive review of the available emergency sanitation technologies, and developing a new MCA framework for sanitation technology selection in emergency relief. First, the paper introduces the challenges in emergency sanitation, then presents a review of the applied and potential emergency technologies, and finally presents a new technology selection framework based on multi-criteria analysis. This review is considered timely and appropriate in light of the increasing complexity and intensity of natural disasters that require more scientific than the traditional approach to emergency sanitation. It also complements the review on emergency water supply (Loo et al., 2012), thus providing a holistic pair of references concerning the emergency water, sanitation and hygiene (WASH) sector.

2.2   Challenges in emergency sanitation

Despite all the invaluable efforts made by the humanitarian agencies in emergency response, the provision of adequate sanitation in emergencies remains a challenge. A review of the past emergencies identified the following key challenges concerning sanitation provision.

Limited access to sanitation infrastructure and resources: Disasters often cause destruction to the existing sanitation facilities and other infrastructure. For instance, extensive damage to sewerage and water supply infrastructure was reported after the Haiti earthquake in 2010, Japanese earthquake and tsunami in 2011, and the typhoon Haiyan in Philippines in 2013 (OCHA, 2011; Oxfam, 2011a; WASH Cluster Philippines, 2014). With such destruction and the often disruptions of the road access to the affected areas, smaller and lighter emergency sanitation relief kits such as packet toilets (e.g. peepoo bags), bucket toilets, and portable toilets (e.g. portaloos), etc., may be the only option available. Water dependent sanitation kits or systems may not be feasible where water supply systems have been destroyed and not rehabilitated. Power breakdown is also a likely consequence of a disaster which may prevent the use of energy-dependent sanitation technologies (Loo et al., 2012). Communication by means of mobile telephony is also often down during emergencies which makes it difficult to reach people that may be of strategic importance, especially in coordinating the sanitation activities. Due to the often chaotic circumstances during emergencies, resources such as trained operators and other technical material, including machinery and vehicles may not be readily available resulting in improper operation and maintenance for a sanitation technology in place. Some of these factors are emergency-specific and some are further amplified in an emergency setting, in combination making the emergency sanitation provision extremely complex and challenging (Brdjanovic et al., 2015). The humanitarian agencies may address such deficiencies by training standby staff for deployment to the emergency cases with acute personnel shortage. Furthermore, they can create reserve emergency stocks of technical materials with characteristics that anticipate the above deficiencies to ensure rapid deployment on a need basis.

Complex emergency scenarios in relation to sanitation: Disaster scenarios are confronted with difficult circumstances such as floods and hard to excavate grounds. Such situations challenge the traditional approach to emergency sanitation that greatly uses the conventional pit latrine and/or its variations (Harvey et al., 2002; von Münch et al., 2006). Furthermore, the recent case in Haiti in 2010 (urban disaster) where land owners prohibited digging of pits/trenches for emergency latrines (Oxfam, 2011a) is a clear manifestation of difficulties in the future of emergency sanitation response given the rise in urban disaster events. Furthermore, often limited (or lack of) vacant space, especially in the urban disaster scenarios presents additional difficulties in providing sanitation alternatives and more so the use of pit latrine (Harvey, 2007; Johannessen, 2011; Johannessen et al., 2011). Limited or lack of waste treatment and disposal facilities in the disaster areas also complicates the emergency sanitation response. A systematic approach in the technology selection process is essential to facilitate the identification of the most appropriate sanitation options in such complex situations.

Technological gaps: Lack of technologies that are adequately tailored to adapt to the disaster scenarios is among the key challenges in emergency sanitation provision (Johannessen, 2011). This has been attributed to lack of innovation concerning the emergency sanitation technologies (Brdjanovic et al., 2015), due to insufficient resource allocation (Johannessen, 2011), and a low interest of the scientific community. Sanitation is often less prioritized and receives less funding than other humanitarian interventions, particularly the water supply (Oxfam, 1996; Fenner et al., 2007). The traditional approach that uses of pit latrines appears is still mostly used although other, more advanced technologies, are available. The application of these alternative technologies may be facilitated by the use of an objective approach in the selection process.

Generally, sanitation provision in the currently changing emergency scenarios is becoming increasingly complex. The challenges discussed above require a departure from the traditional response practices. Research and innovation is necessary to come up with technologies that adequately fill in the existent gaps and at the same time correspond to the changing trends of the emergency scenarios. There is also need to improve on the sanitation technology selection processes by coming up with planning tools that integrate the various components of the emergency sanitation chain. Furthermore, there is need to train and maintain a roster of emergency sanitation response personnel to address shortages during the surge periods.

2.3 Sanitation options for faecal sludge and wastewater disposal in emergencies

This section provides a review of potential emergency sanitation technologies, broadly classified into onsite and offsite technologies. An overview of their suitability with respect to the three globally adopted emergency phases, namely; (i) phase one, of duration up to two weeks, (ii) phase two, of up to six months duration, and (iii) phase three, lasting in excess of six months (Brdjanovic et al., 2015) is presented in Table 2-1 below and further discussed in Sections 2.31 and 2.3.2. Furthermore, a summary of the strengths and weaknesses of the onsite technology options is provided in Table 2-2 and Table 2-3, while those of the offsite options are provided in and Table 2-4.

Table 2-1. An overview of technology options with their suitability at different emergency phases (adapted from Robert et al., 2018)

Technology	Emergency phase		
	Phase 1	Phase 2	Phase 3
Open defecation options			
1. Designated defecation sites	•		
2. Controlled open defecation fields	•		
3. Trench defecation fields	••		
Trench latrines			
1. Shallow trench latrines	••		
2. Deep trench latrine	••	•	
Pit latrines			
1. Shallow family latrines	••		
2. Simple pit latrine	••	••	••
3. Ventilated improved pit (VIP) latrine	•	••	••
4. Borehole latrines	••	•	
Composting latrines			
1. Arborloo	••	••	••
2. Fossa Alterna	••	••	••
3. Urine diverting toilets (UDT)			
Non pit latrines			
1. Pour-flush latrine	••	••	••
2. Overhung latrines	•		
3. Bucket or container latrine	••	•	•
4. Packet latrine	••	•	•
5. Chemical toilet	••		
6. Aqua Privy	•	••	••
7. Storage tank latrine	••	•	
8. Raised toilet	••	••	••
On site sanitation options			
1. Septic tank	•	••	••
2. Anaerobic filter		•	••
3. Anaerobic baffled reactor		•	••
Offsite sanitation systems			
1. Constructed wetlands		•	••
2. Waste stabilization ponds (WSPs)		••	••
3. Trickling filters		•	••
4. Up-flow anaerobic sludge blanket reactor (UASB)		•	••
5. Membrane bioreactors (MBR)		•	••
6. Conventional activated sludge (CAS)		•	•

Suitability of a technology in the different emergency phases is given using bullet points (**two bullet points**: suitable, **one bullet point**: less suitable, **no bullet point**: unsuitable). The level of suitability is determined mainly based on applicability, speed of deployment, start-up and material requirements (Robert et al., 2018).

2.3.1 Onsite sanitation technologies

The onsite sanitation options available for application in emergency relief comprising the dry (suitable for handling faecal sludge only) and the wet systems (for handling wastewater) are presented in Table 2-1 and Table 2-2 respectively, including advantages and limitations, and are discussed further.

Onsite dry sanitation options

Table 2-2. Strengths and weaknesses of key emergency onsite dry sanitation options (Cairncross, 1987; Oxfam, 1996; Davis and Lambert, 2002; Paterson et al., 2007; John Hopkins and IFRC, 2008)

Technology	Advantages	Limitations
Designated defecation sites	No construction activities; inexpensive; rapid deployment; no water or electricity requirements; simple to use; can serve a large population if space is available; can be used in difficult to excavate ground conditions	Requires large space; no privacy; may not be acceptable to many people; not suitable in overcrowded and flood scenarios; odour and flies nuisance; difficult to manage; risk of faecal contamination; risk of surface water contamination; only a temporary solution; better suited to hot, dry climates
Controlled open defecation fields	No massive construction activities; rapid deployment; inexpensive; no water or electricity requirements; easily understood; can serve a large population if space is available; screens provide some level of privacy around the fields; can be used in hard or rocky soils	Requires large space; limited privacy within the fields; may not be acceptable to many people; not suitable in overcrowded and flood scenarios; odour and flies nuisance; risk of faecal contamination; only a temporary solution; difficult to manage; better suited to hot, dry climates
Trench defecation fields	No massive construction activities; rapid deployment; inexpensive; reduces odour and flies nuisance; no water or electricity requirements; easy to use; can serve many people if land is available; no need for faecal sludge treatment as it is covered in-situ	Requires large space, and intensive O&M activities; no adequate privacy; not suitable in overcrowded and flood scenarios; difficult to install in hard or rocky soils; difficult to keep in acceptable hygienic state; only a temporary solution
Shallow trench latrines	Simple; can use local materials; rapid deployment; provides some level of privacy; reduces odour and flies nuisance; easy O&M; easy to use; no water or electricity requirements; faecal sludge contact and handling minimized	Large space requirements; inadequate privacy within the facility; short lifespan; not applicable in flood scenarios; difficult to install in hard or rocky soils; presents odour problems in hot and humid climates
Deep trench latrine	Inexpensive; can use local materials; rapid deployment; simple O&M; reduces odour and flies nuisance; easy to use; no water or electricity requirements; provides adequate privacy; reliable; can serve many people; faecal contamination reduced	Difficult to install in hard to excavate ground conditions; considerable space unless trenches are emptied regularly
Shallow family latrines	Inexpensive, can use locally available materials; provides adequate privacy; rapid deployment; no emptying or treatment	The community must have willingness and ability to install the latrines; large space requirement; not applicable in flood, high water

Technology	Advantages	Limitations
	requirements as faecal sludge is buried; Simple O&M; no water or electricity requirements	table, and difficult to excavate ground conditions
Simple pit latrine	Simple and inexpensive; can use local materials; rapid deployment; low O&M costs; easily understood; easy to use; no water or electricity required; reliable performance and service	Difficult to install in hard to excavate ground conditions; odour and flies nuisance is likely; overflows and infiltration can pollute water sources; large space required to serve many people; may require regular emptying in crowded emergency conditions; emptied sludge may require treatment; emptying costs may be high
Ventilated improved pit (VIP) latrine	Minimal flies and odour nuisance; quality long-term solution; no water and electricity requirements; can use local materials; can serve all types of users (squatters, sitters, wipers, washers); fast start up time; low O&M costs; easily understood; reliable	Relatively costly; relatively long construction time; difficult to construct in non-excavation ground conditions; emptying costs may be significant; emptied sludge may require treatment; requires large space to serve many people
Borehole latrines	Inexpensive; fast installation; can be constructed in hard soils; minimal workforce; fast start up; reliable; no water or electricity requirements; can use local materials; simple O&M; no emptying and sludge handling	High risk of groundwater pollution; odour and flies nuisance likely; easy to block; short lifespan; large space required to serve many people in emergencies; not applicable in flooding conditions or high groundwater table areas
a) *Arborloo*	Simple and inexpensive; can be use local materials; low risk of pathogen transmission; no waste handling	Requires large space due to high fill up rates in emergencies; labour intensive in moving the superstructure; suitable for rural scenarios; not applicable in flood conditions
b) Fossa *Alterna*	Inexpensive, an use local materials; composted waste can be reused; no water and energy requirements; low O&M costs; normally requires less space; long life span	Cover material (e.g. soil, ash, etc.) continuously required; large space required due to high fill up rates in emergencies against short time allowed for waste degradation
c) *Urine diverting toilets (UDT)*	Can be used where infiltration options are viable; no water and energy requirements; offers waste reuse opportunities; simple; inexpensive; can use local materials; low O&M costs; rapid start-up; reliable performance upon proper use and supervision.	Some options e.g. urine diverting dry toilets (UDDT) not applicable if water is used for anal cleansing; requires user awareness for separation and more careful O&M; inadequate decomposition time due to high fill up rates in emergencies may require frequent emptying; may require large space in emergencies; more space required for offsite composting
Pour-flush latrine	Inexpensive; can use local materials; no flies and odour nuisance; easy to deploy, start-up, clean and maintain; good solution where water is used for anal cleansing; no electricity requirements; reliable.	Limited but constant supply of water; can be blocked by solid anal cleansing materials; more costly than the simple pit latrine; may not be applicable in flooding or high water table conditions; effluent and sludge may need additional treatment
Overhung latrines	Inexpensive; fast deployment; an use local materials; may be the only appropriate option in some flooding conditions; no water or electricity requirements; simple O&M; requires less space; can serve many people	Not applicable on stagnant or recreational water bodies or where adverse effects downstream are anticipated ; risk of surface water pollution; odour and flies nuisance; not reliable; a temporary solution only considered once all other options have been explored

Technology	Advantages	Limitations
Bucket or container latrine	Fast deployment; applicable in flood conditions; no water or electricity requirements; requires less space; can provide immediate solution at the onset of the emergency	Containers may be diverted to different uses; may not be acceptable to many people; extensive hygiene education to ensure safe disposal; requires a lot of disinfectant and containers; faecal sludge may require treatment before final disposal; spillages may occur when emptying the buckets
Packet latrine	Can be used in flooded areas; light weight and easy to transport; can be used where space is limited; does not require water or electricity; reliable with good supervision, collection and disposal plan; low cost	Requites prior identification of disposal site; may not be acceptable to many people; generated sludge may need treatment before final disposal; requires constant supply of replacement bag
Chemical toilet (Portaloo)	Easily deployable; hygienic; minimizes odour and flies nuisance; short start-up time; simple to use; does not require constant water or electricity to operate; can be used in flood and non-excavation conditions	Expensive; does not use local materials ; O&M is costly; sludge may require treatment before final disposal
Aqua Privy	Relatively low capital and operation costs; can use local materials; when raised can be used in rocky and flood prone areas; good solution where water is used for anal cleansing; no odour and flies nuisance; small space requirements; no electricity requirements	Requires constant water supply; effluent and sludge quality requires secondary treatment; requires frequent emptying and constant maintenance; long deployment time; not suitable for solid anal cleansing materials
Storage tank latrine	Rapid deployment; suitable for diverse site conditions; containers often available in relief shipments; requires less space as regular emptying of the containers is done; no risk for surface and groundwater contamination; no water or electricity requirements; can be applied in flood or difficult to excavate conditions	Requires regular emptying; requires large number of containers which could be used for other purposes; generated sludge may require treatment before disposal; spills and overflows may pollute the environment
Raised toilets	Rapid deployment; good solution in non-excavation sites; minimizes groundwater contamination; easily understood; no water or electricity requirements; reliable performance and service	Relatively expensive; regular emptying required; requires large space to serve a big population in emergencies; emptying costs may be high compared to capital costs; emptied sludge may require treatment

Open defecation options

The main open defecation options applied in emergency relief include designated defecation sites, controlled open defecation (see Figure 2-1) and trench defecation fields (see Figure 2-2), (Davis and Lambert, 2002; Harvey et al., 2002). The designated defecation site together with its improved version referred as the controlled open defecation field are often provided as the first step in emergency faecal sludge disposal to localize and control indiscriminate defecation.

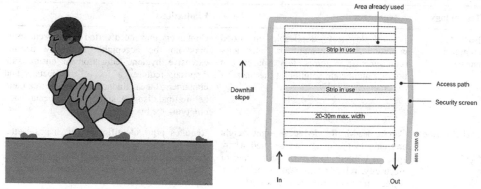

Figure 2-1. Open defecation field (WEDC 2007)

A major characteristic of the controlled open defecation field is the screening to segregate sites for men and women (Davis and Lambert, 2002). This option is suitable in the early emergency stages especially in areas with excavation difficulties. For instance, it was applied in the overcrowded Rwandese refugee camps in Goma, Zaire, in 1994, which were located on a volcanic area with difficulty to dig (Oxfam, 1996). An even more advanced version of open defecation is the trench defecation field in which shallow trenches are dug to defecate in and cover the excreta with soil (Oxfam, 1996; Harvey et al., 2002).

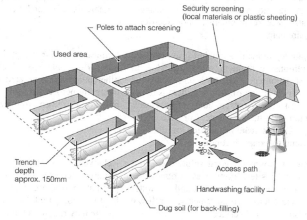

Figure 2-2. Shallow trench defecation field (Harvey et al., 2002 and Reed 2010)

This is important to reduce odour and flies nuisance, and the potential transmission of excreta related diseases. The open defecation options may not be considered as improved sanitation solutions, but may be the only available alternatives in the early emergency stages especially where people are accustomed in open defecation. However, these options lack the aspect of sustainability due to the difficulty to maintain in good hygiene standards (Oxfam, 1996; Davis and Lambert, 2002; Harvey et al., 2002). Hence, their use should be discontinued as soon as better options are available. In fact the use of open defecation should be discouraged wherever possible. Furthermore, the open defecation options require a lot of space which may not be available in urban scenarios thus limiting their application to the rural emergencies.

Trench latrines

The two types of trench latrines include shallow trench and deep trench latrines (see Figure 2-3), which are constructed on the same principle but only differ in their depths. They are easily deployable and commonly applied as fast solutions in emergencies (Davis and Lambert, 2002; Harvey et al., 2002; John Hopkins and IFRC, 2008).

Figure 2-3. Simple trench latrine, Bangladesh (Harvey et al., 2002)

Trench latrines offer a better sanitation solution in emergencies than the open defecation options as they are more hygienic and provide privacy that protects the dignity, especially of the female users. Shallow trench latrines were applied in the emergency after the Pakistani floods in 2010, while deep trench latrines were applied after the Haiti earthquake in 2010 and the Mozambican refugee camps in Malawi in 1998 (Oxfam, 1996; Bastable and Lamb, 2012).

Pit latrines

Several technologies exists that use the pit for faecal sludge containment for which the majority are variations of the conventional pit latrine including the shallow family latrine, simple pit latrine (see Figure 2-4), borehole latrine, ventilated improved pit latrine (VIP) (Davis and Lambert, 2002; Harvey et al., 2002; Paterson et al., 2007), and composting latrines (NWP, 2006; Morgan, 2007; Berger, 2011; SOIL, 2011).

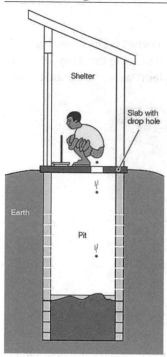

Figure 2-4. Simple pit latrine (WEDC, 2012)

There is no documented evidence of the application of the shallow family latrine in emergencies, but it has a great potential since it can be constructed and maintained by the affected population using locally available materials (WHO, 2009b). This has the advantage to create a sense of ownership leading to proper use and sustainability of the facilities. The simple pit latrine is by far the most popular sanitation option in emergencies (Harvey et al., 2002; von Münch et al., 2006). Its application is however greatly hampered in complexity in disaster scenarios such as floods and urban conditions. The VIP latrine can also provide faecal sludge disposal solution in emergencies with ability to eliminate odour and flies nuisance that are likely in the other pit latrine options. There is no documented evidence of VIP application in the past emergencies, because it is relatively expensive and requires more time to build, hence the inexpensive simple pit latrine is chosen. Another possible pit based technology is the borehole latrines which is suitable in emergencies where a large number of latrines is immediately needed. However, its adoption greatly depends on the availability of the necessary equipment and the groundwater table (Davis and Lambert, 2002). Variants of the composting toilet technology, some of which have been applied in the past emergencies, are available. The main variations include the arborloo, fossa alterna, and the urine diverting toilet (UDT). The use of the arborloo and fossa alterna has not been reported in emergencies, but UDTs were used following the Haiti earthquake in 2010 (SOIL, 2011) and the Bolivian floods in 2008 (Oxfam, 2009).

The majority of the pit latrine technologies are best applied from the 2nd phase of the emergency, but the borehole latrine option can be applied in the early emergency stages. Furthermore, based on their characteristics, the pit latrine technologies are mostly applicable in the rural rather than urban scenarios. Some options however, such as the UDDT can be applied in urban scenarios if secondary composting sites can be availed (SOIL, 2011).

Non-pit latrines

The latrine technologies that do not directly use pits for faecal sludge containment, some of which have been applied in emergencies, include the pour-flush latrines, overhung latrines (see Figure 2-5), bucket or container latrines, packet latrines, chemical toilets, aqua privies (see Figure 2-6), storage tank latrines, the raised toilets (see Figure 2-7) (Cairncross, 1987; Harvey et al., 2002; Paterson et al., 2007; John Hopkins and IFRC, 2008), and the recently developed eSOS smart Toilet (see Figure 2-8) (Brdjanovic et al., 2015).

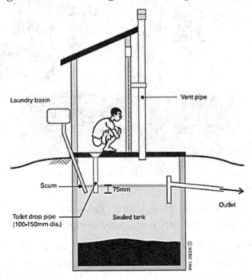

Figure 2-5. Overhung latrine, Bangladesh (Harvey et. al 2002)

Figure 2-6. Aqua Privy (Harvey et. al 2002)

Figure 2-7. Raised toilet units in Haiti (Oxfam, 2011b)

The eSOS smart toilet (Figure 2.8) is a recent addition in the potential technology options for emergency sanitation applications. The toilet is of a urine diversion type with separate collection (and treatment) of urine and faeces, with both 'dry' and 'wet' sanitation options (Brdjanovic et al., 2015).

Figure 2-8. A pilot-scale eSOS smart toilet (Photo. D. Brdjanovic)

A prototype unit has been successfully tested in Tacloban, Philippines (Zakaria et al., 2017) while more rigorous field test of the pilot-scale unit is currently under way in Nairobi, Kenya.

The pour-flush latrine can offer a good solution for faecal sludge disposal especially in emergency scenarios where peopled use water for anal cleansing. For instance, it was used after the Indonesian tsunami in 2004 (Oxfam International, 2008), and the floods in Thailand in 2011 (Dietvorst, 2011). There is often water shortage in emergencies, so the pour-flush water demand can be addressed by installing the latrines side by side with showers and wash rooms to collect grey water for flushing. Overhung latrine is also a potential technology in emergencies but is considered not environmentally sustainable. However, it can be the only possible immediate option in the early emergency stages in extreme conditions such as floods (Harvey et al., 2002; John Hopkins and IFRC, 2008). When necessary, the overhung latrine should only be adopted temporarily as better solutions are sought. Bucket or container latrines (see Figure 2-9) can be applied in emergency scenarios with space constraints. However, the containers should be emptied at least daily and disinfectants applied to minimize contamination and odour (Harvey et al., 2002). As this option involves close contact with faecal sludge, it should only be used as a temporally solution when acceptable to the users.

Figure 2-9. A bucket latrine (https://www.primalsurvivor.net/bucket-toilet/)

Another option that can offer an immediate solution in areas with space constraints, or where better solutions cannot be immediately availed is the packet latrine that uses regular plastic bags or special plastics e.g. the PeePoo bags (see Figure 2-10).

Figure 2-10. Packet latrine (peepoo bag) (http://www.peepoople.com/)

The packet latrines were successfully trialled in the emergency camps after the Haiti earthquake in 2010 (Patel et al., 2011). A chemical toilet or portaloo (see Figure 2-11) is also a possible technology for emergency sanitation that was used during the Kosovo refugee crisis in 1999, the floods in the Dominican Republic in 2003 (Harvey et al., 2002; Harvey, 2007), the emergencies in South and Central America (WHO, 2009b), the Haiti earthquake in 2010 (Oxfam, 2011b), and for Syrian refugees in Turkey in 2013 (UNHCR, 2014). However, its application in emergencies is reported to be expensive with very high operation and maintenance costs (Eyrard, 2011; Oxfam, 2011b; Bastable and Lamb, 2012).

Figure 2-11. Chemical toilets (Portaloos) in Turkey (UNHCR 2013)

Aqua privy is a possible technology that is more suited in emergency scenarios where people use water for anal cleansing. There is no documented evidence on its use in emergencies, but it can be applied in similar conditions as the pour-flush latrines. Storage tank latrines can be used in flood scenarios or where digging is not possible. Final disposal sites and proper emptying mechanisms should be identified in advance as the tanks require regular emptying. Another technology suitable in areas with excavation difficulties is the raised toilet. For instance, they were used in the Mozambican refugee camps in Malawi in 1998 (Oxfam, 1996) and the emergency camps in Haiti in 2010 (Oxfam, 2011a; Bastable and Lamb, 2012).

Generally, the majority of the non-pit latrines are best applied as temporal solutions either in the early emergency stages and should be replaced with better solutions as soon as possible. However, some options such as the pour-flush latrine, aqua privy and raised latrine can be used as long-term interventions from the 2nd phase of the emergency. The locations for options such as overhung toilets need careful consideration to minimize public health risks from the contaminated watercourses. Raw sludge is expected from the majority of the non-pit latrine options, thus a proper management plan should be arranged alongside their selection.

Onsite wastewater disposal options

The main onsite wastewater treatment technologies with potential for application in emergencies include the septic tanks (Cairncross, 1987; Franceys et al., 1992; Harvey et al., 2002; Fenner et al., 2007), anaerobic filters (AF) (Jawed and Tare, 2000), and anaerobic baffled reactors (ABR) (Parkinson and Tayler, 2003; Foxon et al., 2004) as presented in Table 2-2.

Table 2-3. Strengths and weaknesses of key emergency onsite water-based disposal options (Cairncross, 1987; Franceys et al., 1992; Jawed and Tare, 2000, Harvey et al., 2002; Parkinson and Tayler, 2003; Foxon et al., 2004; Fenner et al., 2007)

Technology	Advantages	Limitations
Septic tank	Low capital and operation costs; simple construction and O&M; can use local materials; long service life; reduces odour and flies nuisance; reduces sludge volume; relatively small space requirement; no electricity requirements; reliable if well operated and	Relatively expensive; slow start-up; requires water to operate; effluent and sludge require secondary treatment; conventional septic tank is not suitable for areas with high water table and frequent flooding

Technology	Advantages	Limitations
	maintained; packaged units can be used for easy deployment	
Anaerobic filter (AF)	Can be constructed with local materials; no electricity required; long service life; reduces odour and flies nuisance; reasonable reduction of BOD and solids; resistant to loading variations; low-cost operation; produces stable and less sludge; simple O&M; compact with less space requirement; reliable if well supervised; can be packaged for easy deployment; can serve many people.	Requires water to operate; effluent requires secondary treatment to remove pathogens and nutrients; long start-up time of 6 to 9 months; requires expert design and construction; susceptible to clogging thus restricted to treat wastewater with low solids content; possibilities for odour generation; operates well in warm climates.
Anaerobic baffled reactor (ABR)	Can be constructed with locally available materials; long service life; reduced odour and flies nuisance; high organics reduction; low capital and operation costs; no electricity required; low sludge yield; high resistance to loading variations; minimal maintenance; can serve many people; less space requirements	Long deployment time; requires constant supply of water; pre-treatment required to avoid clogging; requires expert design; effluent and sludge require secondary treatment

A modified form of the conventional septic tank, popularly known as the Oxfam sanitation unit, was used in Bangladesh emergency camps in 1970s. The first units were successfully trialled in a cholera research laboratory in Dacca, Bangladesh, in 1974, after which more units were deployed in the country in 1975-1977 for use in the Bengali refugee camps (Daniel and Lloyd, 1980; Howard, 1996). AF is also an option that can be used to treat highly concentrated wastewater in emergency camps, but its use in past emergencies is not reported. It can be applied by connecting with wet toilet systems or laundry yards, kitchen sinks, and showers. ABR is yet another option for onsite emergency wastewater treatment and was successfully used (in a serial combination with septic tanks and reed beds) in Myanmarese refugee camps in Bangladesh for the treatment of faecal sludge (29 m^3/day) from pour-flush latrines (Porteaud, 2012).

The onsite wastewater treatment technologies may take a relatively long time to put in use when constructed onsite making them applicable onwards from the 2nd phase of emergency. However, their design can allow for packaged plants which can be deployed in the 1st phase of the emergency.

2.3.2 Offsite sanitation technologies

Various offsite sanitation technology options have been applied for both wastewater and faecal sludge management. These technologies are mostly applied in wastewater treatment but, faecal sludge is occasionally applied in co-treatment. Offsite sanitation technologies that can be applied, mainly for wastewater treatment in emergencies include the constructed wetlands (CW) (Kadlec, 2009), waste stabilization ponds (WSP) (von Sperling and Chernicharo, 2005; Mara, 2008), trickling filters (Qasim, 1999; Metcalf and Eddy, 2003; von Sperling and Chernicharo, 2005), up-flow sludge blanket reactor (UASB) (Seghezzo et al., 1998; Chong et al., 2012), membrane bioreactors (MBR) (Judd, 2006; Le-Clech et al., 2006; Williams and Pirbazari, 2007) and the conventional activated sludge systems (CAS) (Qasim, 1999; Metcalf and Eddy, 2003). They are presented in Table 2-2, including advantages and limitations, and are discussed further. CW treatment system has three variations including free-water surface, horizontal subsurface flow, and vertical flow wetlands (von Sperling and Chernicharo, 2005; Kadlec, 2009; Kadlec and Wallace, 2009). There is no reported evidence on application of constructed wetlands in disaster situations but the successful use in regular conditions demonstrate their suitability in the 3rd phase of emergency when more permanent and sustainable sanitation solutions are required. WSP is another technology that is cheap and has great potential for emergency wastewater applications. The WSP system can achieve substantial levels of organic matter, nutrients, and pathogen removal (von Sperling, 2005; von Sperling and Chernicharo, 2005; Olukanni and Ducoste, 2011). The conventional WSP system can be applied in disaster areas where land is available and excavation is possible. Modified units, for instance, using bladders can also be used in areas where excavation is not possible. The application of WSP in emergencies is not well documented, but it is reported that Oxfam used the technology at least three times in emergencies (Johannessen et al., 2011). Deployment of WSPs can take a long time making it suitable for application from the 2nd phase of the emergency. Trickling filters can also be applied at the later stages of the emergencies to provide permanent wastewater treatment solutions. Based on the principle of trickling filters, package plants can also be developed for rapid deployment in the early stages of emergencies. Although the trickling filter technology is widely applied in regular conditions, so far there is no documented evidence on its application in emergencies. UASB is a well-established technology in the large scale industrial wastewater treatment applications (Seghezzo et al., 1998; Ekama and Wentzel, 2008; Chong et al., 2012). It has also been applied in treating low strength domestic wastewater in warm climates (von Sperling and Chernicharo, 2005). So far there is no documented evidence of UASB applications in emergency relief, but the technology provides for package plants that can be rapidly deployed in emergencies. UASB systems can be applied from the 2nd and particularly 3rd phase of the emergency. MBR produces reliable and high quality effluent (Judd, 2006) which can be reused to augment water supplies in the emergency camps. Its success in treating high strength industrial wastewater (Yang et al., 2006; Hoinkis et al., 2012) demonstrates the potential for application in emergencies. Containerized MBRs are available that can be rapidly deployed in disaster areas making it potentially applicable from the 1st phase in emergencies. In 2005, a design proposal for an emergency MBR system was done for Oxfam (Paul, 2005); however, it is not clear whether it was ever implemented. The CAS system has wide application in domestic and industrial wastewater treatment and thus can potentially be applied in emergencies. So far, its application in emergency relief is not reported. Nevertheless, the technology can be applied in the 2nd and particularly the 3rd phase of emergency.

Table 2-4. Strengths and weaknesses of key offsite sanitation technologies (Qasim, 1999; Metcalf and Eddy, 2003; von Sperling, 2005; von Sperling and Chernicharo, 2005; Judd, 2006; Olukanni and Ducoste, 2011)

Technology	Advantages	Limitations
Constructed wetlands (CWs)	High BOD and solids reduction; aesthetically pleasant; no electricity required; can be constructed with locally available materials; no flies and odour problems if used properly; can serve many people; simple construction and O&M practice; no sludge generation; low operational cost	Long deployment time; large space requirement; expert design/supervision; moderate capital cost based on land and liner costs; may provide breeding site for mosquitoes; not applicable in flooding conditions; operation requires continuous supply of water

Technology	Advantages	Limitations
Waste stabilization ponds (WSPs)	Construction can use locally available materials; no electricity required; no significant problems with flies/odors if designed and operated well; simple construction, O&M; low operation cost; satisfactory BOD removal; resistant to loading variations; very low sludge production; can serve many people	Large land area and excavation requirement; may require geological investigations which may not be possible in emergencies, slowing deployment; requires expert design/supervision; variable capital costs depends on the land cost; requires water to run; not applicable in flood conditions; slow start-up; performance dependent on climatic conditions
Trickling filters	Resistant to organic and hydraulic loading rate variations; high BOD removal; simple design and operation; can treat various wastewater streams; reliable with supervised operation; stable and less sludge; can serve many people	Expensive, expert design/construction required, requires constant source of water and electricity; flies and odour issues; pre-treatment required; complex engineering in dosing system; all parts and materials may not be locally available; may take time to deploy in emergencies; large space requirement; dependent on warm climate; not applicable in flood conditions; slow start-up
Up-flow anaerobic sludge blanket reactor (UASB)	Can withstand high organic and hydraulic loading rates; high organic reduction; stable and less sludge; biogas production which is energy source; less space requirement; less energy consumption; reliable with good supervision; slow startup can be accelerated by seeding; can treat grey and black water; can be packaged for easy deployment; can serve many people	Employs complex expert design; requires constant source of water; capital cost may be high since materials for construction may not be locally available; O&M may be difficult due to loading variations likely in emergencies; potential to generate odour; process is sensitive to toxic compounds
Membrane bioreactors (MBR)	Compact, small footprint; resistant to organic and hydraulic loading variations; high quality effluent; high organic matter removal efficiency; can be packaged for easy and fast deployment in emergencies; applicable in complex site conditions; stable, less sludge; reliable if well supervised; can treat multiple wastewater streams; can serve many people; startup can be aided by seeding	High capital and operational costs; requires expert design/construction; cannot be constructed with locally available materials; requires high skilled operation; requires constant source of water and energy to operate; may present environmental problems with noise
Conventional activated sludge (CAS)	Stable and reliable performance; high removal efficiencies of up to 99% in BOD and pathogens achieving a relatively high effluent quality; can be modified to meet specific discharge limits; low land requirements; can be sized to serve many people; can be packaged for easy deployment;	Requires expert design, constant source of electricity and water; high capital and operation costs; materials for construction may not be locally available; excess waste sludge may require additional treatment; high energy consumption; sophisticated operation requires skilled operators

2.3.3 Overview of emergency sanitation practices

Although many disasters have occurred in the past, some are not adequately reported especially concerning emergency sanitation. Disasters resulting in high loss of lives and destruction of property like the Asian tsunami in 2004 (Oxfam International, 2008) and Haiti earthquake in 2010 (Oxfam, 2011a, 2011b) received substantial global attention. Unfortunately, and confirming the lower interest in sanitation than the other elements of emergency response, only a limited number of these disasters are documented with specific information on sanitation (Table 2-5).

Table 2-5. Reported sanitation options in major emergencies

Location	Cause of disaster	Estimated number of people affected [*10^6]	Year	Applied emergency sanitation technologies	Reference
Thailand	Floods	1.5	2011	Packet toilets, pour-flush floating toilets	(Dietvorst, 2011; IDMC/NRC, 2012)
Japan	Earthquake & tsunami	0.5	2011	Toilets and sewage system	(OCHA, 2011)
Port-au-Prince, Haiti	Earthquake	1.5	2010	Pit latrines, portaloo/chemical toilets, UD toilets, packet toilets (biodegradable bags), trench latrines, raised toilets, septic tanks	(Johannessen, 2011; Oxfam, 2011a, 2011b; Bastable and Lamb, 2012)
Uganda	Landslide	0.005	2010	Pit latrines	(Atuyambe et al., 2011)
Pakistan	Floods	20	2010	Open defecation, pit latrines, shallow trench latrines	(UNICEF, 2010; Bastable and Lamb, 2012)
Philippines	Floods	5	2009	Raised latrines	(WHO, 2009a; Bastable and Lamb, 2012)
Pakistan	Earthquake	3.5	2005	Open defecation, pit latrines, shallow trench latrines	(ERRA, 2007; Amin and Han, 2009)
Banda Aceh, Indonesia	Tsunami	2.2	2004	Pour-flush toilets, septic tanks/DEWATS	(Oxfam International, 2008; Johannessen, 2011)
Sri Lanka	Tsunami	2	2004	Portaloo/chemical toilets, pour-flush toilets, septic tanks	(Yamada et al., 2006; Fernando et al., 2009)
Malawi	Mozambican refugee camps on banks of river Shire	0.78	From 1989	Trench latrines, raised family latrines	(Babu and Hassan, 1995; Oxfam, 1996)
Rwandese refugee camps in Goma, DR Congo	Refugees due to Genocide in Rwanda	0.5-0.8	1994	Controlled open defecation zones	(Goma Epidemiology Group, 1995; Oxfam, 1996)
Dadaab, Kenya	Somalia refugees: war/famine	0.3	Since 1992	Pit latrines	(CARE, 2012; Ogol, 2013)
Kosovo	War/Refugee crisis	0.85	1998 - 1999	Portaloo/chemical toilets, deep-trench latrines	(CNN, 1999; Independent International Commission on Kosovo, 2000; Harvey et al., 2002)
DRC Congo	Civil strife/war	1.4	Since 1990s	Pit latrines	(USAID/OFDA, 2009)
Darfur, Sudan	War	1	Since 2003	Pit latrines	(Depoortere et al., 2004; Zakaria, 2013)
Myanmar camps in Bangladesh camps	Civil war	0.029	Since 1990s	Septic tanks, anaerobic baffled reactors	(Porteaud, 2012)

The review in Section 2.3.1 and 2.3.2 above shows existence of a wide range of sanitation technologies, but Table 2-5 reveals that in essence nine technologies were applied in the sixteen past emergencies that were reviewed. Most of the solutions applied are dry technology options for faecal sludge containment being variations of the pit latrine, which traditionally is the most common practice in emergencies (Harvey et al., 2002; von Münch et al., 2006). The continued application of such technologies, even in situations where they obviously are not feasible, implies that it is not only their adequacy that influences the selection process but also other factors such as market forces and political considerations. Particularly, the traditional reliance on proven suppliers by humanitarian agencies may be a major factor influencing the selection. It appears that the reliance and application of the pit latrine gained popularity when emergencies were less complex or less understood making the technology sufficient in many incidences. This may also be the reason why a lot of current developments on the emergency sanitation kits are based on the pit latrine technology. However, this trend is likely to change due to the increasing number of emergencies occurring in urban settings, making the use of pit latrines less adequate. Hence, a diversification of the applied emergency sanitation technologies is likely in the near future as more alternatives are sought, possibly by adapting the conventional technologies to the changing disaster scenarios. Furthermore, the overview on the available literature regarding sanitation in past emergencies does not state clearly how wastewater (sewage) in the majority of emergencies (especially from the care centres) was dealt with. This deficiency is further confirmed in the Sphere Handbook (The Sphere Association, 2018), which does not provide the standards for management of wastewater. Nevertheless, emergency wastewater management can be addressed with easy to deploy package plants derived from the majority of the regular wastewater treatment technologies.

Generally, the deficiencies discussed above suggest that a systematic multi-objective approach is in essence lacking in the emergency sanitation technology selection process, and decisions seem to mostly rely on traditional practices based on empirical approaches. This may limit the decision makers' freedom to explore the applicability of alternative technologies which potentially might also be (even more) suitable to emergency sanitation. The situation can be addressed by use of an appropriate multi criteria framework as a guide in the selection process. The several general sanitation selection frameworks available (Palaniappan et al., 2008) only address sanitation in a development context, which definitely differs from the emergency context. For instance, sanitation provision in emergencies is usually dynamic to conform to the changes during the different emergency phases, which is not the case in the development context where conditions are quite stable. In addition, time and land space constraints are more evident in emergencies. Therefore, a new framework is constituted here (see Section 2.4) based on diverse technology options and assessment criteria, which is potentially applicable to wide spectra of emergencies. The criteria considered include site specific considerations relevant to sanitation provision such as the phase of emergency, number of affected people, water availability, groundwater level, possibility for excavation, energy or power availability, road accessibility, space availability, anal cleansing material, and sludge handling facilities. Also critical aspects related to technology characteristics are considered such as the capital costs, ease of deployment, space requirement, O&M costs and practices, start-up time, use of local materials, and skill requirement. Details of the identified criteria are presented and discussed in the subsequent sections.

2.4 Selection of sanitation technologies in emergency response

Due to the diversity of scenarios in different disasters, it is impractical to adopt one technology for all disasters. The selection process of the emergency sanitation technologies should depend on the type of disaster event, area-specific conditions (Fenner et al., 2007), and the technology characteristics. Hence, it is a complex task requiring a well-structured methodology that considers and applies all important criteria fairly across the available options. In Section 2.4.1 and 2.4.2 the main criteria relating to the local conditions and the technical aspects of the sanitation technologies are established and discussed. Subsequently, a methodology for emergency sanitation technology selection based on compensatory multi-criteria analysis is developed and discussed in Section 2.4.3.

2.4.1 Area-specific related criteria to consider in emergency sanitation technology selection

In order to ensure that the chosen sanitation technology respects the local (prevailing) conditions in the disaster area, it is necessary to apply the area-specific criteria in the selection process. Fenner et al. (2007) suggested a set of site specific criteria for technology selection in emergencies. However, their criteria missed important aspects of wastewater-disposal as they were only considered in the selection process for the faecal sludge-disposal options. Hence, a new set of criteria is proposed and discussed below, that considers all emergency sanitation options. It includes the phase of emergency, number of affected people, water availability, groundwater level, possibility for excavation, energy or power availability, road accessibility, space availability, anal cleansing material, and sludge handling facilities.

The three emergency phases adopted in the emergency sanitation response are defined in Section 2.3. The aim of the response in the first phase of the emergency is to contain the spread of diseases mostly using temporary solutions, while in the second phase (semi) permanent solutions are supplied with aim to reduce morbidity and mortality rates, and any further spread of disease. Permanent solutions are normally supplied in the third phase with the aim to sustain health and wellbeing of the people (Davis and Lambert, 2002; Harvey et al., 2002; Brdjanovic et al., 2015). Based on this specificity, it follows that all technologies cannot be adequately applied in every emergency phase.

The number of affected people is also major factor to consider in the selection process. It should include both the current and the future population based on the projections over the expected emergency duration. The number of the people hosted in an emergency facility determines the faecal sludge and/or wastewater quantities generated, and hence it influences the selection process so that the capacity of the chosen technology can adequately match the amount of waste produced. The choice between offsite and onsite options is also influenced by the number of people. For example, considering land constraints in the emergency settings, authors in this study suggest a minimum of 1,000 persons to justify the adoption of offsite sanitation systems. Nevertheless, the choice of the offsite sanitation systems will also depend on the water availability, which is another important factor influencing the selection process. Water availability in the disaster area will also influence the choice between wet sanitation systems (likely where water is sufficiently available) and dry sanitation systems (more feasible in water scarce conditions) (Fenner et al., 2007).

Groundwater level should also be considered during technology selection to avoid the risk of contaminating groundwater sources, especially for the technologies that require excavation. The Sphere Association (2018) recommends specifically for the pit latrines and soak-away pits, to be located at least 30 m away from any groundwater source and the bottom of the pit to be at least 1.5 m above the groundwater table. The groundwater level becomes increasingly important in disasters related to floods or combined with heavy rainfall events resulting in extremely high water levels which excludes pit latrines. Another criterion in technology selection that is related to the groundwater level is the possibility for excavation. The trench and pit latrines require excavation and hence are dependent on the soil types and the (hydro) geology of a particular area. Excavation may be difficult or impossible in extreme conditions of collapsible, rocky and hard soils, flood conditions (e.g. during the floods in Manila, Philippines in 2009, and some parts of Pakistan in 2010), and land ownership issues (e.g. in Haiti during the earthquake in 2010) (Bastable and Lamb, 2012).

Disasters often cause destruction of infrastructure including power supplies and road networks which are vital for emergency response. Hence, it is essential to consider the power or energy availability and the road accessibility among the criteria for emergency sanitation technology selection. Power breakdown prevents the use of energy-dependent sanitation technologies. Alternatively, energy can be produced from mobile electric generators (Loo et al., 2012). The compatibility of the intended technology/accessories with the available power sources should also be assessed as it may prevent their application. In situations where road accessibility to the disaster site is not possible, alternative means of access such as air transportation are required, in which case the size and weight of the sanitation technology to be availed can be limited. Road accessibility is also important for the sanitation options requiring frequent emptying of large quantities of sludge (Loetscher and Keller, 2002). The rapid fill up of the emergency facilities e.g. pits and vaults may require more frequent emptying by trucks. Hence, assessment of the access to the possible disposal sites is vital as it will determine the type and functioning of the technology mobilized.

The selection process should also consider space availability as there is variation in space requirements among the sanitation technologies. Emergency camps, especially in urban areas are often constructed around existing permanent settlement areas which are usually already congested (Paul, 2005). Harvey (2007) identified lack of space, especially in urban or crowded areas as a major challenge in providing the emergency sanitation. Hence, where land space is limited, technologies with high land requirements such as the CW and WSP may be less feasible. Land space may also be limited by ownership regulations, for instance, in Haiti in 2010 (Oxfam, 2010).

Anal cleansing material is also an important criterion in technology selection as it determines the user interface of the sanitation facility, the fill up rates of the containment vaults, the emptying methods, and the ultimate disposal of waste. The common anal cleansing materials include water, soft material (e.g. tissue paper), or hard material (e.g. stones, wood, corn cobs, etc.) (Harvey, 2007). The use of water for anal cleansing, for instance, may affect the functioning of dry sanitation options e.g. the UDDT, while pour-flush toilets may not be applicable where the hard materials are used.

The assessment of the availability of sludge handling facilities is essential to ensure that the solutions chosen are properly aligned with the sludge management capabilities on ground. Rapid fill up of the sanitation facilities (e.g. the pit and storage tank latrines) is anticipated, especially in the early emergency stages when a few sanitation facilities are available (Brown et al., 2012). Often such facilities are emptied leading to generation of huge volumes of raw sludge that needs proper handling to protect public health. Conversely, there are technologies, e.g. the WSP and MBR, which generate little amount of sludge which is relatively stable. Others do not produce sludge at all, e.g. the fossa alterna and arborloo.

2.4.2 Technology related criteria to consider in emergency sanitation technology selection

Fenner et al. (2007) suggested that the criteria set should include deployability, affordability, space requirements, installation and start up, process stability, operation and maintenance (O&M) skill requirements, operator attention criticality, energy consumption, vector attraction, coliform removal, effluent quality, sludge quantity, sludge quality, noise, odour, and use of local materials. However, some aspects of these criteria such as noise, odour, vector attraction, and coliform removal may not be of critical importance in an emergency context provided that proper O&M of the proposed technologies is ensured. Furthermore, the operator attention criticality can be considered as an integral part of the O& M skill requirements. Hence, a new and lean set of criteria is proposed and redefined here considering the critical technology aspects. The proposed criteria are presented and discussed below which includes the capital costs, ease of deployment, space requirement, O&M costs and practices, start-up time, use of local materials, and skill requirement.

The capital costs include the costs for buying, delivering, and installing the technology. Cost assessment of the most notable components e.g. land, groundwork, electromechanical equipment, and construction (Veenstra et al., 1997) is critical. The criticality of the capital cost in the technology selection in emergency context may differ from the development context. For example, technologies with high costs may be more acceptable for emergency sanitation (problem driven) than in the regular sanitation (demand driven).

The ease of deployment is also a crucial criterion (Fenner et al., 2007; Loo et al., 2012) that relates to the simplicity and speed of availing the facility and putting it in operation at the point of use. Some technologies are bulky and heavy, so they require assembly on site; others are compact and light, and may be deployed to the site readily assembled. The ease of deployment is most critical at the early emergency stages when solutions should be rapidly provided to safeguard public health.

Another crucial criterion is the space requirement (Fenner et al., 2007) which accounts for the total space required by a sanitation facility with its accessories. This can greatly influence the technology choice, especially in situations where land is highly limited such as in the urban scenarios. For instance, more compact technologies such as the MBR would be more preferable in crowded scenarios than the WSP or CW that require large space.

The selection process should also consider O&M costs and practices to ensure that the installed facility offers proper service. Assessment should be conducted to ascertain the requirements (e.g. the maintenance frequency, complexity, and downtime, etc.), and the access to the resources required (e.g. availability and costs of spare parts, chemicals/reagents, energy, etc.) for the day to day running of the system (Peter-Varbanets et al., 2009).

Another key criterion that will influence the technology selection is the start-up time related to the speed and time taken by a system (after the installation) to realize the intended functioning. Long start-up time may be experienced with the biological systems such as the UASB reactors (von Sperling and Chernicharo, 2005). Conversely, simpler options such as pit latrines are used immediately after the installation.

Use of local materials is also an important aspect that will influence the technology selection. When the facility can be constructed with locally available materials, a significant reduction in the implementation time and costs may be realized.

Skill requirement is also a crucial factor in the emergency technology selection (Fenner et al., 2007; Loo et al., 2012). If a technology requires skilled, semi-skilled, or unskilled personnel must be clear, as well as the level of training required to operate. Since resources (including skilled personnel) are often limited in emergency situations, technologies with relatively simple operation requirements are generally more desirable.

To be able to quantify the technology performance concerning the above set of criteria, each criterion can be related to score values to provide input to the decision-making system. Hence, each criterion was assigned a qualitative score on a scale from 1-5, from the least desirable to the most desirable, respectively. Table 2-6 gives an overview of the definition of the scores in relation to the evaluation criteria.

Table 2-6. Criteria used for evaluation of emergency sanitation technologies and definition of the scores applied

Evaluation criteria	Definition of scores				
	1	2	3	4	5
Capital costs	Very high cost	High cost	Moderate cost	Low cost	Very low cost
Ease of deployment	Very big and heavy; can only be assembled on site and requires heavy handling equipment/machinery	Big and heavy; can only be assembled on site but does not require heavy handling equipment/machinery	Some parts can be assembled off site which can be eventually joined to make the facility at the site	Relatively small and light; can be assembled off site and transported to the site	Very simple to set up at the site e.g. simple pit latrine
Space requirement	Very high land area requirement	High land area requirement	Moderate land area requirement	Low land area requirement	Very low land area requirement
O&M costs	Complicated maintenance; done regularly; time-consuming	Complicated maintenance; done regularly; not time-consuming	Lightly complicated activities; done regularly; not time-consuming	Simple maintenance; done occasionally; not time-consuming	No significant maintenance required
Start-up time	Long start up time; more than 1 week, even with the use of seeding	Long start up time, more than a week; with no use of seeding	Moderate start up time less than a week; but with use of seeding	Short start up time (approximately 3 days); but with use of seeding	Very short start up time (less than a day); Can be used immediately after installation
Use of local materials	Construction or operation of the unit/facility entirely uses proprietary/imported materials or accessories	Construction of the unit/facility requires part locally available material, and part imported/proprietary materials and relies on imported/proprietary accessories to operate	Construction of the unit/facility requires part locally available material, and part imported/proprietary materials but does not rely on imported/proprietary accessories to operate	Construction of the unit/facility uses entirely locally available materials but uses imported/proprietary accessories to operate	Construction of the unit/facility and accessories entirely uses locally available materials (no imported/proprietary materials)

Evaluation criteria	Definition of scores				
	1	2	3	4	5
Skill requirements	Very high complexity; requires very high level skills and can only be handled by very well trained personnel. User needs intensive training to ensure proper use of the facility	High level of complexity; can only be handled by highly skilled personnel and the user needs some training to use the facility properly	Moderate complexity; requires moderate level of technical knowledge and can be handled by semi-skilled personnel and the user can utilize the facility with simple demonstration	Low level of complexity; simple technology can be handled by unskilled personnel but the user needs some simple demonstration to the use of the facility	Very low level of complexity; simple technology that requires no or limited technical knowledge to operate and use. User can utilize the facility without difficulty

The scores are later applied for technology evaluation in Section 2.4.3.

2.4.3 Methodology for technology selection process

Variations of the local conditions in the disaster areas make each emergency situation rather unique. In addition, the available emergency sanitation technologies exhibit a broad range of features. As presented in Section 2.4.1 and 2.4.2, there are various contending factors that influence the selection of the most appropriate sanitation technology. Therefore, a structured selection process that considers the various contending factors is essential. In this section, a methodology for the emergency sanitation technology selection process is developed. Figure 2-12 summarizes the steps involved in the selection process in the form of a flowchart. Two key stages are performed including screening and evaluation as discussed in Section 2.4.3 respectively.

Screening stage

The screening stage comprises Steps 1 and 2 (see Figure 2-12) in which the potentially feasible technologies are assessed against the site specific criteria provided in Section 2.4.1 to identify a smaller set of feasible technologies. The technically infeasible technologies are eliminated leaving only those which are applicable under the given prevailing conditions. For instance, acute water scarcity in a particular area eliminates the possibility to adopt wet sanitation technologies such as pour-flush toilets and aqua privies. Similarly, pour-flush toilets are unsuitable where hard materials e.g. corn cobs are used for anal cleansing. Table 2-7 illustrates the screening process in which the feasible options are marked in white, while the non-feasible options are marked in grey.

Table 2-7. Decision table for identification of feasible emergency sanitation technologies

No.	Criteria / Technologies	Phase of emergency — 1st	Phase of emergency — 2nd	Phase of emergency — 3rd	Number of affected people — <1000	Number of affected people — >1000	Water availability — Limited	Water availability — Adequate	Level of groundwater — <2m	Level of groundwater — 2-5m	Level of groundwater — >5m	Possibility for excavation — Excavate-able	Possibility for excavation — Non excavateable	Energy/power availability — Available	Energy/power availability — Not available	Road accessibility — No accessibility	Road accessibility — Limited accessibility	Road accessibility — Full accessibility	Space availability — Limited	Space availability — Adequate	Anal cleansing material used — Water	Anal cleansing material used — Soft	Anal cleansing material used — Hard	Availability of sludge handling facilities — Available/can be availed	Availability of sludge handling facilities — Not available
1	Designated defecation sites		■																■						
2	Controlled open defecation fields																		■						
3	Trench defecation fields								■										■						
4	Shallow trench latrine																								
5	Deep trench latrines								■																
6	Shallow family latrine																								
7	Simple pit latrine								■	■															■
8	Ventilated improved pit latrine								■	■															■
9	Borehole latrine								■																■
10	Compositing latrines								■								■						■		■
11	Pour-flush latrine	■														■									
12	Overhung latrine		■																						■
13	Bucket latrine		■	■																					■
14	Packet latrine																								■
15	Chemical toilet																							■	
16	Aqua privy	■		■				■																■	
17	Storage tank latrine																							■	
18	Raised latrine																							■	
19	Septic tank	■					■										■								
20	Anaerobic filter	■					■																		
21	Anaerobic baffled reactor	■					■																		
22	Constructed wetlands	■											■				■								
23	Waste stabilization ponds	■											■										■		
24	Trickling filter	■																					■		
25	UASB reactor	■																							
26	Membrane biorector	■																							
27	Conventional activated sludge	■																							

Legend

	Recommendable
■	Non-recommendable

The resulting lean set of the feasible options minimizes the effort of having to deal with many options in the ensuing evaluation stage.

Evaluation stage

At the evaluation stage all the options that pass the screening stage are assessed against the technology related criteria provided in Section 2.4.2. Ultimately, the evaluation results for all the options are compared with each other to support the decision making. This stage comprises eight steps starting from step 3 in Figure 2-12.

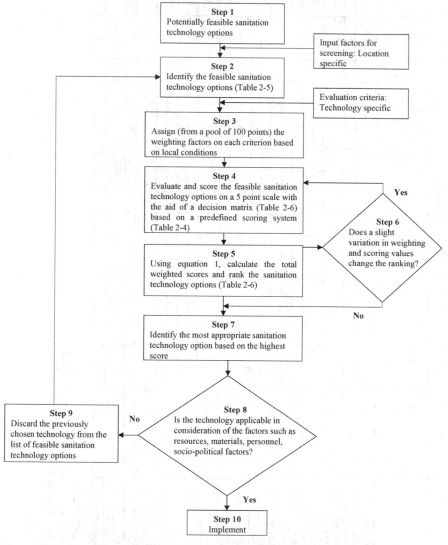

Figure 2-12. Flow chart summarizing emergency sanitation technology selection process (adapted from Loo et al., 2012)

In step 3, a weighting factor is assigned to each criterion to adjust its relative importance based on the needs of affected people and the prevailing conditions in the disaster area. The weighting values are assigned by distributing 100 points among the criteria based on their relevant significance in the particular situation. In this paper, for instance, criteria such as space requirement, start-up time, and deployability were considered to be critical determinants for technology suitability and were assigned a weight factor of 20. The criteria which depict recurrent activities and involve costs such as O&M were assigned a weight factor of 15. Criteria that were considered important but can be mitigated during the design and implementation phase such as the operator skill requirements, and those which require one-off expenditure and are likely to be accepted at whichever range in emergencies, such as the capital costs were assigned a weight factor of 10. Finally, the criteria which were considered important but of less priority such as the use of local materials were assigned a weight factor of 5. Hence, the total points assigned to the seven criteria above sums up to 100. However, it is important to note that these are only default demonstration values and that the weight assigned to a particular criterion is dependent on the prevailing conditions on the ground so the decision makers can vary the weights according to their priorities.

In Step 4 the feasible technologies are scored against the evaluation criteria based on the scoring system provided in Table 2-6. Step 5 involves compiling the total weighted scores of each technology and ranking them by means of the decision matrix (Table 2-8).

Table 2-8. Decision matrix for emergency sanitation technology comparison with an example of scoring and weighting

No.	Criteria	Capital costs	Ease of deployment	Space requirement	O&M costs	Start-up time	Use of local materials	Skill requirements	Scores	
	Weight factor	10	20	20	15	20	5	10	Nominal/35	Weighted /500
1	Dedicated designated sites	5	5	1	4	5	5	5	30	405
2	Controlled open defecation fields	5	5	1	4	5	5	5	30	405
3	Trench defecation fields	5	4	1	5	5	5	5	30	400
4	Shallow trench latrine	5	5	2	2	5	5	5	29	395
5	Deep trench latrines	5	4	3	5	5	5	5	32	440
6	Shallow family latrine	5	4	1	4	5	5	5	29	385
7	Simple pit latrine	4	4	2	5	5	5	5	30	410
8	Ventilated improved pit latrine	4	4	1	4	5	4	5	27	370
9	Borehole latrine	4	4	2	4	5	4	5	28	390
10	Composting latrines	5	4	1	5	5	5	5	30	400
11	Pour-flush toilet	4	3	1	4	5	5	5	27	355
12	Overhung latrine	5	5	3	5	5	5	5	33	460
13	Bucket toilet	5	5	3	2	5	3	5	28	405
14	Packet toilet	5	5	4	2	5	1	5	27	415
15	Chemical toilet	4	4	3	1	5	1	5	23	350
16	Aqua privies	4	3	3	3	5	3	4	25	360
17	Storage tank latrine	4	4	3	2	5	3	5	26	375
18	Raised toilet	5	4	1	3	5	3	5	26	360
19	Septic tank	4	4	3	4	5	3	4	27	395
20	Anaerobic filter	4	4	3	4	3	4	4	26	360
21	Anaerobic baffled reactor	4	4	3	4	3	4	4	26	360
22	Constructed wetlands	4	1	1	5	4	4	3	22	285
23	Waste stabilization ponds	4	1	1	5	4	4	4	23	295
24	Trickling filter	4	3	3	4	3	2	3	22	320
25	UASB reactor	2	3	4	3	2	1	2	17	270
26	MBR reactor	1	4	5	1	5	1	1	18	320
27	Conventional activated sludge	3	3	4	3	3	1	2	19	300

The weighted total score of a technology is determined according to the equation below.

$$S_i = \sum_{j=1}^{n} w_j \cdot x_{ij} \qquad i = 1,2,3,4 \dots m \qquad \text{(Eq 2-1)}$$

Where S_i is the total weighted score, w_j and x_{ij} are the weighting factor and score of the ith sanitation technology for the jth criterion, respectively while m represents the last option in a range of one to m^{th} sanitation options.

Step 6 involves the performance of a sensitivity analysis done by introducing small variations in both the scores (e.g. 1 point) and weighting factors (e.g. ≤ 3) to assess their effect on the ranking of the options. In case a big variation is observed, the ranking and the calculation of weighted scores (steps 4 to 6) are repeated until a sensible ranking is obtained.

Step 7 identifies the technologies with the highest score (best options). If there is a clear 'winner' (margin between the winner and the runner-up is significant, e.g. arguably considered to be 5-10% of the total score) the process can proceed to step 8. In case that two or more technologies are very close to each other the process should be either repeated reconsidering the weighting and scoring applied or continue to Step 8 with two or three technologies.

Step 8 re-assesses the availability of resources such as funds, materials, and personnel for the ease of implementation (step 10) of the selected sanitation option. In case incurable limitations are observed in relation to the considerations in step 8, the chosen technology is discarded (step 9) from the list of feasible options and the cycle is repeated.

The final weighted scores for the technologies presented in the decision matrix (Table 2-8) are partially subjective. While the first step in the (nominal) scoring is based on the more objective technology related criteria, the second step in the (weighted) scoring is based on the weighting factors derived from an ideal emergency scenario. In principle therefore, the nominal scores for each technology as presented in Table 2-8 are not expected to change concerning a real case application. This is because the scores are based on the criteria derived from the technology characteristics that are fixed. On the other hand, changes may occur in the corresponding final weighted scores which are influenced by the variable weighting values depending on the needs of the affected people and the prevailing conditions in the disaster area. The significance of a particular criterion may change from one emergency context to another as well as may change by the time depending on the priorities and the local conditions. This accordingly affects the assigned weighting values. It can thus be inferred that the final weighted scores presented here (Table 2-8) are quite subjective and can only be used to demonstrate the proposed methodology. Nevertheless, from the decision matrix it is obvious that one-size-fits-all approach cannot be applied in emergency sanitation. Hence, a systematic choice of the sanitation technologies to be deployed for emergency response is essential. The MCA developed here can create a departure from the traditional approach to emergency sanitation which relies on standard remedies and intuition of the planners. It is expected that proposed MCA (that uses an inclusive scientific based approach) will significantly address the deficiencies resulting from the current approach in sanitation technology selection for emergency response. In addition, the application of the MCA may bring more technologies into practice, thus motivating the manufacturers and technology suppliers to further explore, develop and promote technologies for emergency sanitation. Furthermore, future innovations in emergency sanitation are anticipated, thus the proposed MCA provides flexibility to add new technologies in the matrix whenever they become available.

2.5 Conclusions and recommendations

A review of sanitation technology options and their applicability in emergency sanitation provision, and an evaluation on the selection processes applied in sanitation technology selection during emergencies was carried out in this study. The study revealed the existence of a wide range of onsite and offsite sanitation technologies with a high potential to be applied in emergency sanitation provision. However, based on an overview of the past emergency cases, it was observed that only a few of those technologies were actually applied in disaster relief, majority of which are based on the conventional pit latrine and/or its variations. Furthermore, it was observed that the lack of a suitable framework for sanitation technology selection in emergencies is causing the replication of standard solutions even in varied scenarios leading to often sub-optimal service delivery. The selection processes is mainly influenced by the standard practices of the individual humanitarian actors as well as the intuition of the planners. However, it is observed that the selection process of the emergency sanitation technologies should depend on the type of disaster event, area-specific conditions, and the technology characteristics. Therefore, a MCA which is developed in this paper provides a systematic tool for selecting the most appropriate sanitation technology for a given emergency scenario to ensure better services. Finally, it is desirable to stimulate dialogue, exchange of information and views between stakeholders usually involved in the emergency field, but also stimulate the other groups of interest (such as manufacturers, sanitation experts, consultants, academia, authorities and governments, etc.,) to join and actively participate in development of novel and practical approaches to emergency sanitation.

References

Amin, M.T., Han, M.Y., 2009. Water environmental and sanitation status in disaster relief of Pakistan's 2005 earthquake. Desalination 248, 436-445.

Atuyambe, L.M., Ediau, M., Orach, C.G., Musenero, M., Bazeyo, W., 2011. Land slide disaster in eastern Uganda: rapid assessment of water, sanitation and hygiene situation in Bulucheke camp, Bududa district. Environmental Health 10.

Babu, S.C., Hassan, R., 1995. International migration and environmental degradation—The case of Mozambican refugees and forest resources in Malawi. Journal of Environmental Management 43, 233-247.

Bastable, A., Lamb, J., 2012. Innovative designs and approaches in sanitation when responding to challenging and complex humanitarian contexts in urban areas. Waterlines 31, 67-82.

Berger, W., 2011. Technology review of composting toilets: Basic overview of composting toilets (with and without urine diversion). GIZ GmbH, Eschborn Germany. Retrieved from http://www.susana.org/docs_ccbk/susana_download/2-878-2-1383-gtz2011-en-technology-review-composting-toilets1.pdf.

Brdjanovic, D., Zakaria, F., Mawioo, P.M., Garcia, H.A., Hooijmans, C.M., Ćurko, J., Thye, Y.P., Setiadi, T., 2015. eSOS® – emergency Sanitation Operation System. Journal of Water Sanitation and Hygiene for Development 5, 156-164.

Brown, J., Cavill, S., Cumming, O., Jeandron, A., 2012. Water, sanitation, and hygiene in emergencies: summary review and recommendations for further research. Waterlines 31, 11-29.

Cairncross, S., 1987. Low-Cost Sanitation technology for the control of Intestinal Helminths. Parasitology Today 3, 94-98.

CARE, 2012. Update: Horn of Africa Food Security Emergency, Retrieved from http://www.care.org/sites/default/files/documents/EMER-2012-Horn-of-Africa-emergency-one-year-report.pdf.

Chong, S., Sen, T.K., Kayaalp, A., Ang, H.M., 2012. The performance enhancements of upflow anaerobic sludge blanket (UASB) reactors for domestic sludge treatment – A State-of-the-art review. Water Res. 46, 3434-3470.

CNN, 1999. NATO struggling to keep refugee camps sanitary., Retrieved from http://edition.cnn.com/WORLD/europe/9904/07/kosovo.refugees.02/.

Daniel, R.R., Lloyd, B.J., 1980. Microbiological studies on two Oxfam sanitation units operating in Bengali refugee camps. Water Res. 14, pp 1567-1571.

Davis, J., Lambert, R., 2002. Engineering in emergencies: A practical guide for relief workers, 2nd ed. ITDG Publishing 103–105 Southampton Row, London WC1B 4HL, UK.

Depoortere, E., Checchi, F., Broillet, F., Gerstl, S., Minetti, A., Gayraud, O., Briet, V., Pahl, J., Defourny, I., Tatay, M., Brown, V., 2004. Violence and mortality in West Darfur, Sudan (2003–04): epidemiological evidence from four surveys. Lancet 2004; 364: 1315–20.

Dietvorst, 2011. Sanitation Updates: News, Opinions and Resources for Sanitation for All. Thailand. floating toilets for flood-hit areas, Retrieved from http://sanitationupdates.wordpress.com/2011/11/03/thailand-floating-toilets-for-flood-hit-areas/.

Ekama, G.A., Wentzel, M.C., 2008. Organic Material Removal, in: Henze, M., van Loosdrecht, M.C.M., Ekama, G.A., Brdjanovic, D. (Eds.), Biological Wastewater Treatment: Principles, Modelling and Design. IWA Publishing, London, UK., p. 526.

ERRA, 2007. Fountains of Life: Rebuilding Water and Sanitation Systems in Earthquake-affected Areas in Pakistan., Pakistan. Retrieved from http://www.erra.pk/Reports/KMC/CaseStudies/WATSAN.pdf.

Eyrard, J., 2011. Portable toilets in emergencies: lessons learned from Port-au-Prince, Haiti, 35th WEDC International Conference. WEDC, Loughborough, UK.

Fenner, R.A., Guthrie, P.M., Piano, E., 2007. Process selection for sanitation systems and wastewater treatment in refugee camps during disaster-relief situations. Water and Environment Journal 21, 252-264.

Fernando, W.B.G., Gunapala, A.H., Jayantha, W.A., 2009. Water supply and sanitation needs in a disaster – Lessons learned through the tsunami disaster in Sri Lanka. Desalination 248, 14-21.

Finney, B.A., Gearheart, R. A., 1998. Water and Wastewater Treatment Technologies Appropriate for Reuse Model (WAWTTAR).

Foxon, K.M., Pillay, S., Lalbahadur, T., Rodda, N., Holder, F., Buckley, C.A., 2004. The anaerobic baffled reactor (ABR): An appropriate technology for on-site sanitation. Water SA 30, 44-50.

Franceys, R., Pickford, J., Reed, R., 1992. A guide to the development of on-site sanitation. WHO, Geneva, Switzerland.

Goma Epidemiology Group, 1995. Public health impact of Rwandan refugee crisis: what happened in Goma, Zaire, in July, 1994? The Lancet 345, 339-344.

Harvey, P.A., 2007. Excreta Disposal in Emergencies: a Field Manual. Water, Engineering and Development Centre (WEDC), Loughborough University, Leicestershire, UK.

Harvey, P.A., Baghri, S., Reed, R.A., 2002. Emergency Sanitation: Assessment and programme design. WEDC, Loughborough University, UK, Leicestershire, UK., p. 349.

Hoinkis, J., Deowan, S.A., Panten, V., Figoli, A., Huang, R.R., Drioli, E., 2012. Membrane Bioreactor (MBR) Technology – a Promising Approach for Industrial Water Reuse. Procedia Engineering 33, 234-241.

Howard, J., 1996. Rethinking the unthinkable - effective excreta disposal in emergency situations. WATERLINES 15, 5-6.

IDMC/NRC, 2012. Global estimates 2011: People displaced by natural hazard-induced disasters, Retrieved from http://www.internal-displacement.org/8025708F004BE3B1/%28httpInfoFiles%29/1280B6A95F452E9BC 1257A22002DAC12/$file/global-estimates-2011-natural-disasters-jun2012.pdf.

Independent International Commission on Kosovo, 2000. The Kosovo Report Conflict, International Response, Lessons Learned. Oxford University Press.

Jawed, M., Tare, V., 2000. Post-mortem examination and analysis of anaerobic filters. Bioresour. Technol. 72, 75-84.

Johannessen, Å., 2011. Identifying gaps in emergency sanitation, Design of new kits to increase effectiveness in emergencies, 2 day Workshop, 22-23 February 2011, Stoutenburg, The Netherlands, . WASTE and Oxfam GB, Retrieved from http://www.susana.org/images/documents/07-cap-dev/b-conferences/13-stoutenberg-conference-2011/stoutenberg-feb-2011-report-final.pdf.

Johannessen, A., Patinet, J., Carter, W., Lamb, J., 2011. Sustainable sanitation for emergencies and reconstruction situations - Factsheet of Working Group 8, in: (SuSanA)., S.S.A. (Ed.), Retrieved from http://www.susana.org/docs_ccbk/susana_download/2-797-9--wg08-en-susana-factsheet-wg08-emergencies-final-ci-tms-evmx.pdf.

John Hopkins and IFRC, 2008. Water, sanitation and hygiene in emergencies in: John Hopkins and IFRC (Ed.), Public Health Guide for Emergencies, 2nd ed. John Hopkins and IFRC, pp. pp. 372-434.

Judd, S., 2006. The MBR Book: Principles and Applications of Membrane Bioreactors in Water and Wastewater Treatment. . Elsevier, London. .

Kadlec, R.H., 2009. Comparison of free water and horizontal subsurface treatment wetlands. Ecol. Eng. 35, 159-174.

Kadlec, R.H., Wallace, S.D., 2009. Treatment wetlands, 2nd ed. Taylor & Francis Group., New York.

Le-Clech, P., Chen, V., Fane, T.A.G., 2006. Fouling in membrane bioreactors used in wastewater treatment. Journal of Membrane Science 284, 17-53.

Loetscher, T., Keller, J., 2002. A decision support system for selecting sanitation systems in developing countries. Socio-Economic Planning Sciences 36, 267-290.

Loo, S.-L., Fane, A.G., Krantz, W.B., Lim, T.-T., 2012. Emergency water supply: A review of potential technologies and selection criteria. Water Res. 46, 3125-3151.

Mara, D., 2008. Waste Stabilization Ponds: A Highly Appropriate Wastewater Treatment Technology for Mediterranean Countries, in: Baz, I., Otterpohl, R., Wendland, C. (Eds.), Efficient Management of Wastewater : Its Treatment and Reuse in Water Scarce Countries. Springer Berlin Heidelberg, pp. 113-123.

Metcalf, Eddy, 2003. Wastewater Engineering: Treatment and Reuse, 4th ed. McGraw-Hill Publishers New York, NY. .

Morgan, P., 2007. Toilets that make compost: low-cost, sanitary toilets that produce valuable compost for crops in an African context. , Stockholm Environment Institute, EcoSanRes

Programme, Stockholm, Sweden. , Retrieved from http://www.ecosanres.org/pdf_files/ToiletsThatMakeCompost.pdf.

NWP, 2006. Smart Sanitation Solutions: Examples of innovative, low-cost technologies for toilets, collection, transportation, treatment and use of sanitation products. KIT Publishers, Amsterdam, Netherlands.

OCHA, 2011. Japan • Earthquake & Tsunami Situation Report No. 1 6 1 April 2011. Situation Report.

Ogol, G.T., 2013. Personal communication, UNESCO-IHE Institute for Water Education, Delft, Netherlands.

Olukanni, D.O., Ducoste, J.J., 2011. Optimization of waste stabilization pond design for developing nations using computational fluid dynamics. Ecol. Eng. 37, 1878-1888.

Oxfam, 1996. Sanitation in emergency situations: Proceedings of an International Workshop, held in Oxford, December 1995, in: Adams, J. (Ed.), Oxfam working paper. . Oxfam (UK and Ireland), Oxford, UK.

Oxfam, 2009. UD Toilets and Composting Toilets in Emergency Settings., Oxford, UK. Retrieved from http://policy-practice.oxfam.org.uk/publications/ud-toilets-and-composting-toilets-in-emergency-settings-126692.

Oxfam, 2010. Haiti Progress Report 2010, Oxford, UK. Retrieved from http://policy-practice.oxfam.org.uk/publications/haiti-progress-report-2010-133955.

Oxfam, 2011a. Haiti Progress Report January-December 2011, Oxford, UK. Retrieved from http://policy-practice.oxfam.org.uk/publications/haiti-progress-report-january-december-2011-200732.

Oxfam, 2011b. Urban WASH Lessons Learned from Post-Earthquake Response in Haiti, Oxford, UK., pp. 11. Retrieved from http://policy-practice.oxfam.org.uk/publications/urban-wash-lessons-learned-from-post-earthquake-response-in-haiti-136538.

Oxfam International, 2008. Tsunami Fund: End of Program Report December 2008, Retrieved from http://www.oxfam.org/en/policy/tsunami-fund-end-program-report.

Palaniappan, M., Lang, M., Gleick, P.H., 2008. A Review of Decision-Making Support Tools in the Water, Sanitation, and Hygiene Sector. Pacific Institute: Environmental Change and Security Program (ECSP), California, USA. , pp. 96. URL http://www.pacinst.org/wp-content/uploads/2013/2002/WASH_decisionmaking_tools2013.pdf.

Parkinson, J., Tayler, K., 2003. Decentralized wastewater management in peri-urban areas in low-income countries. Environment and Urbanization, 15 (1) (2003), pp. 75-90 15, 75-90.

Patel, D., Brooks, N., Bastable, A., 2011. Excreta disposal in emergencies: Bag and Peepoo trials with internally displaced people in Port-au-Prince. Waterlines 30, 61-77.

Paterson, C., Mara, D., Curtis, T., 2007. Pro-poor sanitation technologies. Geoforum 38, 901-907.

Paul, P., 2005. Proposals for a rapidly deployable emergency sanitation treatment system, 31st WEDC International Conference, Kampala, Uganda

Peter-Varbanets, M., Zurbrügg, C., Swartz, C., Pronk, W., 2009. Decentralized systems for potable water and the potential of membrane technology. Water Res. 43, 245-265.

Porteaud, D., 2012. Emergency sanitation workshop Delft: Challenges of waswater disposal. Retrieved from http://www.susana.org/images/documents/07-cap-dev/d-workshops/Delft-2012/3-1_Problem_Definition-UNHCR_Porteau.pdf.

Qasim, S.R., 1999. Watewater treatment plants: Planning, Design, and Operation, 2nd ed. CRC Press, Boca Raton, USA.

Seghezzo, L., Zeeman, G., van Lier, J.B., Hamelers, H.V.M., Lettinga, G., 1998. A review: The anaerobic treatment of sewage in UASB and EGSB reactors. Bioresour. Technol. 65, 175-190.

SOIL, 2011. The SOIL Guide to Ecological Sanitation. Sustainable Organic Integrated Livelihoods (SOIL), Sherburne NY, USA. Retrieved from http://www.oursoil.org/wp-content/uploads/2012/01/2-SOIL-Guide-to-EcoSan-Toilets.pdf, p. 144.

Tappero, J.W., Tauxe, R.V., 2011. Lessons Learned during Public Health Response to Cholera Epidemic in Haiti and the Dominican Republic. Emerging Inefectious Diseases 17, 2087–2093.

The Sphere Association, 2018. The Sphere Handbook: Humanitarian Charter and Minimum Standards in Humanitarian Response, 4rd ed. Practical Action Publishing, Rugby, United Kingdom, p. 406.

UNHCR, 2014. UNHCR WASH Manual: Tools and Guidance for Refugee Settings.

UNICEF, 2010. Children in Pakistan Every Child's Right - Responding to the Floods in Pakistan.

USAID/OFDA, 2009. Democratic Republic of the Congo – Complex Emergency Situation Report #1, Fiscal Year (FY) 2009

van Buuren, J.C.L., 2010. Sanitation Choice Involving Stakeholders: A Participatory multi-criteria method for drainage and sanitation system selection in developing cities applied in Ho Chi Minh City, Vietnam. Wageningen University, the Netherlands. Accessed from http://edepot.wur.nl/157236.

Veenstra, S., Alaerts, G.J., Bijlsma, M., 1997. Chapter 3 - Technology Selection, in: Helmer, R., Hespanhol, I. (Eds.), Water Pollution Control - A Guide to the Use of Water Quality Management Principles. WHO/UNEP.

von Münch, E., Amy, G., Fesselet, J., F, 2006. Ecosan Can Provide Sustainable Sanitation in Emergency Situations with Benefits for the Millennium Development Goals. Water Practice & Technology © IWA Publishing 1.

von Sperling, M., 2005. Modelling of coliform removal in 186 facultative and maturation ponds around the world. Water Res. 39, 5261-5273.

von Sperling, M., Chernicharo, C.A.L., 2005. Biological Wastewater Treatment in Warm Climate Regions. IWA Publishing, London. Seattle.

WASH Cluster Philippines, 2014. WASH baseline Barangay assessment typhoon Haiyan, Philippines. Technical report.

Watson, J.T., Gayer, M., Connolly, M.A., 2007. Epidemics after Natural Disasters. Emerging Infectious Diseases • www.cdc.gov/eid Vol. 13, No. 1.

WHO, 2002. Environmental health in emergencies and disasters : A practical guide. WHO, Geneva, Switzerland.

WHO, 2009a. Health Cluster Situation Report No. 17: Tropical Storm Ketsana – Typhoon Parma. WHO, Philippines.

WHO, 2009b. Technical Note for Emergencies:Technical options for excreta disposal in emergencies. URL http://www.who.int/water_sanitation_health/publications/2011/tn14_tech_options_exc reta_en.pdf.

Williams, M.D., Pirbazari, M., 2007. Membrane bioreactor process for removing biodegradable organic matter from water. Water Res. 41, 3880-3893.

Yamada, S., Gunatilake, R.P., Roytman, T.M., Gunatilake, S., Fernando, T., Fernando, L., 2006. The Sri Lanka Tsunami Experience. Disaster Management & Response 4, 38-48.

Yang, W., Cicek, N., Ilg, J., 2006. State-of-the-art of membrane bioreactors: Worldwide research and commercial applications in North America. Journal of Membrane Science 270, 201-211.

Zakaria, F., 2013. Personal communication, UNESCO-IHE Institute for Water Education, Delft, Netherlands.

Zakaria, F., Thye, Y.P., Hooijmans, C.M., Garcia, H.A., Spiegel, A.D., Brdjanovic, D., 2017. User acceptance of the eSOS® Smart Toilet in a temporary settlement in the Philippines. Water Practice and Technology 12, 832.

Zurbrügg, C., Tilley, E., 2007. Evaluation of existing low cost conventional as well as innovative sanitation system and technologies. NETSSAF deliverable D22&23. Swiss Federal Institute of Aquatic Science and Technology (EAWAG), Duebendorf, Switzerland. URL http://www.susana.org/docs_ccbk/susana_download/2-1350-19en-evaluation-existing-low-cost-technologies-2007.pdf.

Chapter 3
Innovative approaches to emergency sanitation: The concept of the emergency sanitation operation system - eSOS®

This chapter is based on:

Brdjanovic, D., Zakaria, F., Mawioo, P.M., Garcia, H.A., Hooijmans, C.M., Ćurko, J., Thye, Y.P., Setiadi, T., 2015. eSOS® – emergency Sanitation Operation System. Journal of Water Sanitation and Hygiene for Development 5, 156-164.

Abstract

This paper presents the innovative emergency Sanitation Operation System (eSOS) concept created to improve the entire emergency sanitation chain and provide decent sanitation to people in need. The eSOS kit is described including its components: eSOS smart toilets, an intelligent faecal sludge collection vehicle-tracking system, a decentralized faecal sludge treatment facility, an emergency sanitation coordination center, and an integrated eSOS communication and management system. The paper further deals with costs and the eSOS business model, its challenges, applicability and relevance. The first application, which took place in the Philippines brought valuable insights on the future of the eSOS smart toilet (Zakaria et al., 2017). It is expected that eSOS will bring changes to traditional disaster relief management.

3.1 Introduction

In general, an emergency can be considered to be the result of a man-made and/or natural disaster, whereby there is a serious, often sudden, threat to the health of the affected community which has great difficulty in coping without external assistance. Emergency sanitation intervention is a means of promoting best management practice in order to create a safer environment and minimize the spread of disease in disaster-affected areas, and of controlling and managing faecal sludge, wastewater, solid waste, medical waste, and dead bodies. In June 2012, an international emergency sanitation conference was hosted by IHE Delft in Delft where more than 200 experts from relief agencies, governments, academia and industry gathered, and discussed emergency faecal sludge management and public health. It was confirmed that (i) emergency-specific sanitation is not at the forefront of the scientific community, (ii) current solutions are in most cases technologically and economically suboptimal, (iii) there is, in general, insufficient communication between key stakeholders, (iv) academia and practitioners are insufficiently involved, (v) emergency sanitation (technological) development is often associated with drivers such as humanitarian aid agencies or the army, (vi) emergency water supply is given much more attention than sanitation, and (vii) the smart innovative emergency sanitation management (and governance) system is lacking. This concept aims to address these deficiencies and provide sustainable, innovative, holistic, and affordable sanitation solutions for emergencies (such as floods, tsunamis, volcano eruptions, earthquakes, wars, etc.) before, during, and after disaster.

3.2 eSOS®

The abbreviation eSOS stands for the innovative 'emergency Sanitation Operation System' concept (Brdjanovic et al., 2013). This concept addresses the entire emergency sanitation chain Figure 3-1.

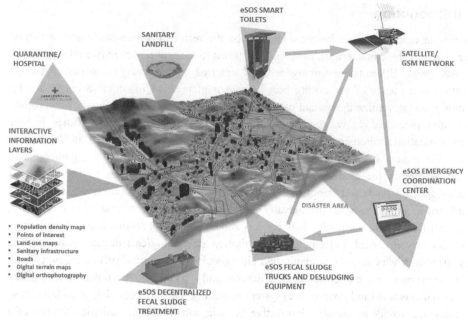

Figure 3-1. eSOS concept components

It is based on a balanced blend of innovative sanitation solutions and existing information technologies adapted to the specific conditions of emergency situations and in informal settlements. The central points of the system are the reinvented smart emergency toilet and the innovative decentralized treatment of faecal sludge, embedded in an intelligent emergency sanitation operation system. Information and communication technologies have a unique opportunity to assist following disasters because the core of any emergency management effort is integration, sharing, communication, and collaboration, things that stakeholders involved embrace and promote.

3.3 eSOS® kit

The eSOS is based on different system components integrated into an easily deployable emergency sanitation kit consisting of hardware and software components. The software components include the communication chain by controlling the mobile network and the Local Area Network (LAN)/Wide Area Network (WAN) simultaneously. The routing application supports receiving data messages – from General Packet Radio Service networks and the SMS channel – from large quantities of Global System for Mobile Communications (GSM) and Code Division Multiple Access (a radio channel access method) units at the same time. Alternatively, a non-GSM-based system can be applied for disaster sites which are not covered by a GSM network (e.g. remote refugee camps) or are temporarily without GSM coverage due to a disastrous natural event. In addition, a portable navigation system is used to supplement faecal sludge collection vehicle tracking. Geographical information system maps and data as well as other interactive and public domain information are used and combined into this integral eSOS, such as digital orthophotography, digital terrain maps, land-use maps, sanitary points of interest, and population density maps. It is all combined in user friendly software with an intuitive graphic interface to allow rapid advance to expert user level. The components of the eSOS are smart toilets, intelligent faecal sludge collection vehicle-tracking systems, decentralized faecal sludge treatment facilities, emergency sanitation coordination centers, and integrated eSOS communication and management systems.

3.4   eSOS® smart toilets

Sanitation facilities usually provided by relief agencies and armies have additional specifications and requirements in comparison to those regularly used in other settings. The smart eSOS toilets have the following characteristics: they are stackable and lightweight, fit a Euro-size pallet, are made of durable materials, are easy to wash and clean, are easy to empty, require minimum maintenance, are raised above the ground, do not require any excavation to install, allow more frequent use, provide excellent value for money, are easy and safe to use, provide privacy, are easily deployable, give a sense of dignity to users, look great and invite usage, etc. Beside these aspects, the eSOS concept addresses the 'smartness' of the emergency sanitation toilet by incorporating unique (either as 'built-in' or 'add-on') features such as: interchangeable squatting pans or sitting toilet, delivered as a urine diversion dry toilet or flush toilet, safe and easy-to-empty storage of urine and faeces, fully solar-powered with up to 7 days energy independency, GSM-based communication, GPS-based tracking, real-time information on occupancy, volume of urine collected, volume of service water and grey water and UV interior disinfection, nano-coated interior, smart card reader entry system, SOS panic button, smart software for monitoring, data collection and optimization, etc. Beside smart data collection and communication, the eSOS toilet is subject to technological innovations from the sanitary engineering perspective. It is a urine diversion toilet with separate collection (and treatment) of urine and faeces, with both 'dry' and 'wet' sanitation options. It is important to note that the eSOS toilet is not designed as an on-site treatment unit due to its high-frequency use and limited storage capacity. The rule of thumb applied by relief agencies of a maximum of 50 users per day will be evaluated during field testing and verified later by data gathered from eSOS toilets to be installed worldwide operating under different conditions. At the moment the capacities of the urine tank and faeces tank in one unit is 200 and 80 L, respectively. This arrangement will be revised after extensive experimental testing. It allows for an emptying interval of individual units of about once a week for a 'dry' toilet. In case the 'community' type of arrangement is applied (several toilets in a cluster), a common larger storage tank will replace individual units allowing for significantly larger storage, more frequent use and less frequent emptying. Longer retention times and ongoing processes in stored faeces and urine will be taken into account in the design of such clustered applications at a later stage of the development of the eSOS system. Of course, the situation will change in the case where continuous or intermittent water supply system and sewer system are available where the 'wet' option may well be applied. As the urine tank makes up part of the toilet body, it will be possible to empty it only on-site by gravity or by a vacuum truck. For faeces evacuation, several emptying options will be possible: by vacuum truck, by replacing a full tank with an empty one, and by several ways of emptying the tank manually on-site (e.g. there is an analogy with vacuum cleaner bags). Owing to specific emergency requirements, its innovative light-weight, stackable toilet structure is proposed to be made of recycled biodegradable materials (like bio-plastic made from potato skins). Options for both on-site and centralized treatment (and their combination)

55

of urine and faecal sludge is also investigated. Packed, a complete toilet kit occupies a volume of $2\ m^3$ which will allow for compact and cheaper shipping (a toilet fits to one standard pallet). Owing to its modular set-up, it will be possible to quickly and simply install the toilet on the spot. Simplified instructions on how to install and use the toilet will be provided with the kit. Each part of the toilet is unique and can only be assembled in one way so as to avoid confusion. In the near future, possibilities to produce toilets locally shall be explored, also using local materials. However, in general it will not be possible to produce these toilets at the disaster location. The present version of the toilet allows for its usage by both children and adults and women and men. In addition, several variations of the eSOS toilet will be produced in a later stage of development to account for different settings and conditions and user groups including elderly citizens, people with disabilities and the injured. Development of the eSOS smart toilet is carried out in two steps, namely: design, manufacturing, and field testing of the 'experimental toilet' (Figure 3-2) and based on the feedback from field testing and relief practitioners, the 'design vision toilet prototype' (Figure 3-3) will be manufactured.

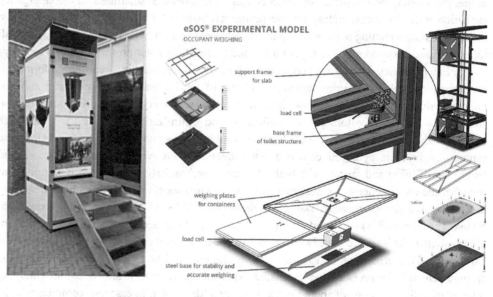

Figure 3-2. eSOS smart experimental toilet awaiting shipment to Philippines for field testing. The toilets' structure and electronic features have been subject to extensive testing during the manufacturing phase (Photo: D. Brdjanovic; Drawings: FLEX/the INNOVATIONLAB)

Figure 3-3. eSOS smart toilet design vision prototype (Photo: D. Brdjanovic)

3.5   Intelligent faecal sludge collection vehicles tracking system

In emergency situations, due to high traffic and load to toilets, frequent emptying (of relatively fresh urine and faeces) is required, which consequently creates demand for well-organized logistics for faecal sludge collection, a feature which is regularly lacking during, by definition, rather chaotic emergency circumstances. As an emergency may last for days, months, and sometimes years, the issue of faecal sludge management and logistics becomes extremely important in sustaining the emergency sanitation chain. For example, in the first few months after the 2010 earthquake disaster in Haiti, the costs for de-sludging toilets and latrines exceeded USD 0.5 Million. The eSOS envisages the use of GPS- (or satellite-) based communication infrastructure; e.g. a real-time GPS vehicle-tracking system, where each truck and/or each trailer/cistern will be equipped with 'easy-to-install' GSM/GPS sensor/card (similar to those supplied with or to eSOS toilets), which will allow 24/7 information of the position (and route) of each toilet-emptying vehicle. This information will 'feed' the advanced, commercially available, vehicle tracking system, and software and on-board location-based analysis, which will process data and provide much useful information (e.g. route optimizer, total amount of urine and faeces collected per day, disposal location, etc.) to the user in the emergency sanitation coordination center. Efforts will be made to create the possibility to rapidly update the navigation maps with the most recent information regarding the disaster event (accessibility of roads, bridges, tunnels, etc.) and isolate sections with limited or no traffic, most likely based on physical site inspection with the eventual support of updated satellite images that can be purchased on demand as an add-on feature of the integrated eSOS.

3.6 Faecal sludge treatment facility

Three distinctive emergency sanitation phases are generally adopted in the work of relief agencies, namely: (i) phase 1 of duration up to 2 weeks, where the main mean for sanitation provision is individual, mass-production, inexpensive kits (like biodegradable PeePoo bags), (ii) phase 2, lasting up to a few months, where substantial sanitation hardware components are supplied to the disaster site (like individual portable toilets or clusters of those, and de-sludging equipment and vehicles), and (iii) phase 3, which can last from several months up to a few years or longer where more (semi)permanent sanitation hardware is supplied such as community based toilets and (mobile) faecal sludge treatment facilities (more sophisticated package/containerized plants or, sometimes, on-site/land-based simplified solutions). Comparatively much higher load (increased usage per toilet), consequent requirements for more frequent emptying, and different faecal sludge characteristics (fresh biologically non-stabilized sludge and fresh non-hydrolyzed urine, with higher public health risk), are distinctive, but often overlooked features of emergency sanitation. Therefore, the current management practices in emergency sanitation need a thorough revision and re-thinking, especially from the treatment perspective, as to this aspect 'the business as usual' approach is applied, often not being fully aware of specific technological and social key issues of concern.

Although many standard options for faecal sludge management in general already exist, their application in emergency situations is not well understood and is often lacking. To address these deficiencies, we shall conceptualize, design, manufacture, test and apply on a pilot scale an innovative, compact, and efficient treatment of emergency sanitation faecal sludge, including (separate) treatment of urine, by physical–chemical treatment-based technologies (e.g. microwave technology and/or dewatering/drying) with specific attention on public health (epidemiologic) aspects and safe disposal of treatment residuals (Figure 3-4).

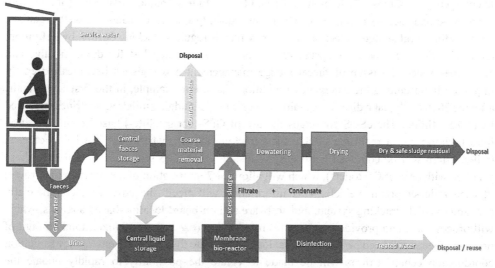

Figure 3-4. eSOS® faecal sludge treatment concept

For the emergency sanitation, also due to economies of scale, it is more appropriate to apply treatment solutions in decentralized on-site settings, rather than solving the faecal sludge issue at the level of individual toilets. However, the decentralized technology being developed here is equally applicable (with some modifications) at small scales as well. After initial testing in the Netherlands, the installation will be subject to field testing in Kenya using excrement collected from the slums of Nairobi.

3.7 Emergency sanitation coordination centre

The emergency sanitation coordination centre is the heart of eSOS concept, and should be located either on-site or at any remote location outside the disaster area. It has a high degree of automation and requires an operator/coordinator. For the immediate response, and if no skilled operator is available, remote operation is possible by an expert operator located outside the disaster area. The coordination centre will be equipped with the central information processing unit (laptop or tablet) which will contain all necessary software and will receive and process all relevant information for the eSOS in the cloud. If on-site, the centre will be responsible for physically inspecting and verifying some of the key information collected by remote sensing and making sure that the correct information is used e.g. accessibility of roads and correct location of existing sanitary infrastructure used in emergency, like sanitary landfill, decentralized wastewater treatment plant, empty industrial storage tanks, and in extreme cases, temporary discharge points to open environment, etc.

3.8 eSOS® operation

Based on information, such as population density maps or real-time population tracking using mobile telephony and other information automatically acquired from the disaster area, in combination with the user-entered information, the operator will have a rather good understanding of where to position the emergency aids. The number of sanitation units deployed will be initially determined using rules of thumb (e.g. up to 50 people per toilet per day), but the application of eSOS will very soon provide practical feedback on these rules as much more (new) data will become available. In addition, based on the existing population density and real-time information on the population migration using mobile telephone signals, the optimal locations (density) of available sanitation units (Pee-Poo bags, for example, for the immediate response, followed by the supply of emergency toilets) will also be determined. In the case of pre-fabricated eSOS toilets, they will automatically report their location to the central system (coordination centre) and will appear on the interactive disaster area map as such. In cases, where the toilet is not equipped with an eSOS kit, it can easily be retrofitted by rapid installation of the necessary sensors and electronic equipment. Also, already existing units can be upgraded with this equipment, so that the entire emergency sanitation facilities are tagged and included in the network. The second step is to equip the faecal sludge collection vehicles with the tracking electronic and navigation equipment. This can be done very easily and quickly by installing the removable equipment preferably inside the driver's cabin. The third step is to mobilize the central data collection and processing unit with all the required software necessary for the operation of the eSOS and to ensure that the Internet connection or access to a cloud computing/server facility via a satellite connection is available. After the system is up and running, the operator can use all above-described features to apply the eSOS in a rapid, more efficient, and economic fashion, with increased confidence. The eSOS system is designed as a stand-alone application, refined at the operator's emergency centre. It enables the definition of the required procedure for each stage in an emergency and to react to every call within the shortest period of time. It also enables the local operator to define the unit's parameters according to both the customer's and local network demands and to create (daily, weekly, monthly) reports with statistics and performance indicators. The authors and funding agencies plan to disseminate all useful feedback from the upcoming practical applications in a separate paper(s) and through other methods of communication. These will soon after be translated into a user manual or operational guide as a part of the eSOS emergency kit.

3.9 Costs and eSOS business model

The current conceptual state of the development of the system does not yet allow being accurate where the costs are concerned for the following reasons. The costs and benefits will depend on many factors where the production and operations costs combined with the location-specific conditions and scale of disaster and number of people affected/served will determine the total financial picture. As both emergencies and disasters have a high degree of uncertainty associated with them and since disasters can strike anywhere in the world at any given moment and given that emergencies have different characteristics and phases, it makes the current application of standard sanitation financial models inadequate and only remotely accurate and useful. Therefore, as a part of the eSOS concept, the development of a holistic business model is demanded and is currently being developed with extended boundaries to capture aspects traditionally difficult to estimate (thus often neglected) but essential to such an assessment, such as costs (and benefits) related to public health (hospitalization, absence from work, productivity, temporary or permanent disability and casualties, quality of life, dignity, safety, etc.). The model will be interactive, adaptable to local conditions and specifics of emergency sanitation, and will also include costs for production (e.g. rotational molds, materials, 'add-ons', labour, etc.), costs for storage, transport and erection, costs for operation and maintenance, and costs for eventual deployment, depreciation, etc. It is expected that in the majority of emergency situations the additional unconventional features and elements of the system and associated costs will be at least compensated for if not overwhelmed by the benefits that such a system can bring. The new eSOS business model will include feedback from major relief agencies and all other key players in emergency relief, will also include demonstrations with detailed costs analyses, and will be verified on several case studies that shall provide more confidence in using it. The business model will be in the public domain.

3.9.1 Challenges

The eSOS confirms the rule that one involved in the process of moving from invention to innovation faces a number of challenges such as how to make a product which will match its purpose at an affordable price with maximized benefits. The eSOS components are designed to satisfy specific requirements of relief operations regarding materials, durability, resistance to theft and misuse, demands of users, environmental and public health, cultural and social features of societies, and must also be attractive to people so that they make use of it in the first place. Expectedly, the eSOS concept cannot possibly be a solution for each and every emergency situation and its future will depend on acceptance, affordability, effectiveness and efficiency of operations, and the extent to which the limitations will be overcome by further development and incorporation of the feedback from practical applications.

3.9.2 Applicability and relevance

The strength of eSOS is that it is addressing, improving, and making each component of the emergency sanitation chain smarter, taking care that innovations also take place at the level of the system. The eSOS system is globally applicable to a wide spectrum of emergency situations where external aid is needed for sanitation. The eSOS concept, with minor adaptations, can be made equally suitable for, but is not limited to (i) sanitation management under challenging conditions usually prevailing in urban-poor areas, such as slums and informal settlements, (ii) sanitation provision for visitors of major open-air events such as concerts, fairs, etc., and (iii) solid waste management. So far, initial constructive and in general encouraging feedback from several parties, including the United Nations Children's Fund (UNICEF), United Nations Refugee Agency (UNHCR), Red Cross, Oxfam, Save the Children, Doctors without Borders (MSF) and OPEC Fund for International Development (OFID), has already been received. It is planned to have key players in relief provision more actively involved in the further development of the eSOS system. Part of the research in the Philippines and other locations will provide us with lessons and answers on how to ensure the uptake of the system. At the moment, the framework for how to commercialize the eSOS and build a business case for the new eSOS enterprise is drafted. It will also include important aspects such as after-sales services that will be very much dependent on the type of emergency, local conditions, culture, emergency setting, etc. The fate of eSOS in a post-disaster period will also be considered. If the life returns to 'normal' and original infrastructure is recovered, the eSOS can be cleaned, dismantled and reused elsewhere as the system allows for it. In the case where new (semi)organized settlements are created, like refugee camps, the eSOS may remain there, given that a proper governance system and the business case are in place to make it sustainable, making the eSOS of more permanent character. In the case, where the eSOS is used for non-emergency situations (events, etc.); it will be reused. In the case of its use in informal settlements (slums), it will be of permanent character. Present design takes care as much as possible that the system is theft proof (the comment on theft and costs of eSOS came up often in social media). The potential clients/end-users are relief agencies, municipalities, water and sewerage companies, solid waste companies, army, police, fire brigades, as well as private sector companies and water supply, and sanitation vendors. The primary goal of eSOS is to save lives by providing an efficient and effective sanitation service during and after emergencies through minimizing the risk to public health of the most vulnerable members of society. The secondary goal is to reduce the investment, operation, and maintenance costs of emergency sanitation facilities and service as a pre-requisite for sustainable solutions, especially in the post-emergency period.

3.10 Concluding remarks

The innovative eSOS concept provides a sustainable, innovative, holistic, and affordable sanitation solution for emergencies before, during, and after disasters. eSOS does not only reinvent the (emergency) toilet and treatment facilities, but uses existing information and communication technology to bring innovation and potential cost savings to the entire sanitation operation and management chain, and most importantly, is expected to improve the quality of life of people in need.

Acknowledgments

The eSOS® concept is developed under the project Stimulating local innovation on sanitation for the urban poor in Sub-Saharan Africa and South-East Asia' financed by the Bill & Melinda Gates Foundation. The eSOS® smart toilet design concept is a joint effort of UNESCO-IHE, FLEX/the INNOVATIONLAB and SYSTECH. The first eSOS® smart toilet testing in the Philippines is supported by the Asian Development Bank and Bill & Melinda Gates Foundation. The eSOS® concept is an invention of UNESCO-IHE Institute for Water Education.

References

Brdjanovic, D., Zakaria, F., Mawioo, P.M., Garcia, H.A., Hooijmans, C.M., Setiadi, T., 2013. eSOS® (emergency Sanitation Operation System): Innovative mergency sanitation concept, 3rd IWA Development Congress & Exhibition, Nairobi, Kenya.

Zakaria, F., Thye, Y.P., Hooijmans, C.M., Garcia, H.A., Spiegel, A.D., Brdjanovic, D., 2017. User acceptance of the eSOS® Smart Toilet in a temporary settlement in the Philippines. Water Practice and Technology 12, 832.

Chapter 4
Microwave technology as a viable sanitation technology option for sludge treatment

4.1 The microwave technology

Microwave is a part of the electromagnetic spectrum with wavelengths (λ) ranging from 1 m to 1 mm and frequencies between 300 MHz ($\lambda =1$m) and 300 GHz ($\lambda =1$mm) (Haque, 1999; Tang et al., 2010; Remya and Lin, 2011). Three microwave frequency bands exist comprising the ultra-high frequency (UHF: 300 MHz ($\lambda=1$ m) to 3 GHz ($\lambda=100$ mm)), the super high frequency (SHF: 3 GHz ($\lambda=100$m) to 30 GHz ($\lambda=10$mm)) and the extremely high frequency (EHF: 30 GHz ($\lambda=10$ mm) to 300 GHz ($\lambda=1$ mm)). Microwaves have higher energy but shorter wavelengths than the radio waves. Conversely, they have lower energy but longer wavelengths than infrared, visible, ultraviolet (UV), X-rays and gamma-rays (Figure 4-1).

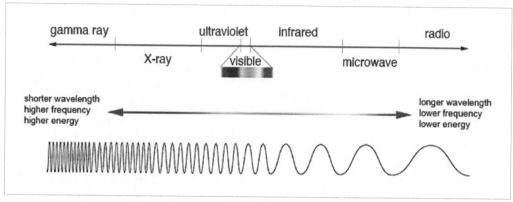

Figure 4-1. The electromagnetic spectrum (NASA, 2016)

Traditionally, the microwave technology was mostly applied in the field of communication, but has now been extensively used in other fields. Some frequencies within the microwave range are used for cellular phones, radar, and television satellite communications. A number of frequencies have also been allocated for applications in the Industrial, Scientific, Medical and Instrumentation (ISMI) fields, with the most popularly applied being 0.915 GHz (approx. $\lambda=328$ mm) and 2.450 GHz (approx. $\lambda=122$ mm) (Bradshaw et al., 1998; Haque, 1999; Thostenson and Chou, 1999). Microwave energy is more expensive than the electrical energy from which it is derived with conversion efficiencies approximately 50% for the 2.450 GHz and 85% for the 0.915 GHz frequency. Microwave irradiation is widely used in heating applications and its unique operational principle offers many advantages over the conventional heating (Haque, 1999; Thostenson and Chou, 1999). For instance, in contrast to the conventional thermal processes where energy is transferred to the material by convection, conduction, and radiation of heat from the surfaces of the material, microwave energy is directly delivered to the materials through molecular interaction with the electromagnetic field. In addition, microwave heating is achieved by conversion of electromagnetic energy to thermal energy (energy transfer) while conventional heating is achieved by heat transfer as a result of energy transfer due to thermal gradients. Since microwaves can penetrate materials and deposit energy at the molecular level, heat can be generated throughout the bulk of the material resulting to rapid and uniform volumetric heating. Furthermore, the microwaves capability to achieve energy transfer at the molecular level can be applied for selective heating of materials. For example, when materials of different dielectric properties are irradiated, microwaves will selectively couple with the

64

higher loss material (Thostenson and Chou, 1999). Other advantages of microwave irradiation over the traditional thermal processes include quick start-up and stopping, and a higher level of safety and automation (Haque, 1999). Microwave technology has been used for heating purposes since 1937, but the first commercial application was found in the 1940s when Percy L. Spencer (an American engineer) invented the domestic microwave oven, which commonly uses 2450 MHz frequency with relatively low power generators (i.e. magnetrons) in the range of 0 - 250 W (Haque, 1999; Eskicioglu et al., 2008; Tang et al., 2010; Remya and Lin, 2011). Conversely, industrial microwave heating systems mostly use 915 MHz frequency with magnetrons as high as 75 kW power.

4.2 Microwave heating system

A microwave heating system comprises four basic components: the power supply, the microwaves source, the transmission lines, and the applicator. The microwave source (vacuum tubes like magnetrons, traveling wave tubes (TWTs), and klystrons) generates the electromagnetic radiation, and the transmission lines (e.g. waveguides) transmit the electromagnetic radiation to the applicator (cavity) which holds the target material (Haque, 1999; Thostenson and Chou, 1999). Electromagnetic radiation results from acceleration of charge. High power and frequencies are required for microwave heating, and that is the reason why vacuum tubes are mostly used as microwave sources. Magnetron tubes, which are used in domestic microwave ovens, are efficient, reliable, and mass produced, and thus are the lowest cost microwaves source available. The magnetron tubes are only capable to generate a fixed frequency electromagnetic field since they use resonant structures for the generation. Conversely, a TWT is used for generating electromagnetic field in a variable frequency microwave since its design allows amplification of a broad band of microwave frequencies in the same tube.

4.2.1 Magnetron

A vacuum tube has the anode at high potential compared to the cathode. As a result of the potential difference, a strong electric field is produced and the cathode heated to remove the loosely bound valence electrons. Once removed from the cathode, the electrons are accelerated toward the anode by the electric field. In a magnetron (Figure 4-2), an external magnet is used to create a magnetic field perpendicular to the electric field. The applied magnetic field creates a circumferential force on the electron as it is accelerated to the anode causing the electron to travel in a spiral direction, thus creating a swirling cloud of electrons. The anode has resonant cavities, so as electrons pass, the cavities resonate and set up oscillations in the electron cloud with a frequency governed by the size of the cavities. Electromagnetic energy (i.e. from the oscillating electron cloud) produced by the resonant cavities is coupled from one of the cavities and channelled through a waveguide to the applicator of a microwave oven or beamed out into the air by an antenna or satellite dish in radar equipment (Thostenson and Chou, 1999).

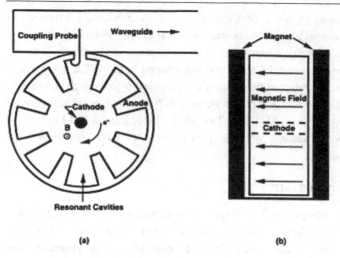

Figure 4-2. Schematic diagram of the magnetron microwave tube: (a) top view, (b) side view (Thostenson and Chou, 1999)

The control of a magnetron tube to achieve an average power output can be done in two common methods: either by adjusting the period of operation or adjusting the cathode current or magnetic strength. For the magnetrons in the domestic microwave ovens which are operated at full power, the average power output is achieved through an on/off type of control, commonly known as duty cycle control in which the cathode current is turned on and off for segments of the period within the specified operation time. Conversely, if a continuous power is required, the average output power of the magnetron can be varied by changing the current amplitude of the cathode or by changing the intensity of the magnetic field (Thostenson and Chou, 1999).

Microwave heating equipment can be batch type similar to the domestic microwave oven or of continuous type which is fitted with a conveyor belt on which materials are moved through the applicator (Figure 4-3) (Haque, 1999).

66

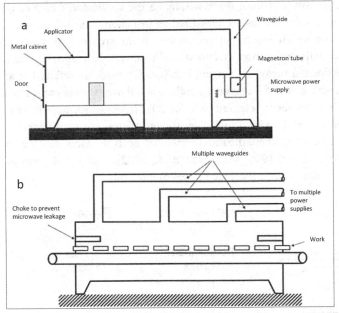

Figure 4-3. a) Batch-type equipment b) Continuous-type equipment (Haque, 1999)

4.3 Principles of microwave technology

The ability of a material to absorb microwave energy and subsequently get heated depends on its dielectric properties: dielectric loss factor (ε'') and dielectric loss constant (ε'). Dissipation factor, also known as tangent loss (tan δ), is the ratio of the dielectric loss factor to the dielectric loss constant of the material and it reflects the dielectric response of materials. It is used to describe the overall efficiency of any material to absorb microwave energy and convert it to heat at a specific frequency (Clark et al., 2000). The dielectric constant (i.e. ratio of the permittivity of a substance to the permittivity of free space) depicts the ability of material to delay or retard microwave energy as it passes through. The loss factor depicts the amount of input microwave energy that is lost by being converted (dissipated) to heat within the material. Hence, materials with high dielectric loss factors are easily heated by microwave energy. The complex relative permittivity (ε^*), which is the fundamental electrical property that relates the material's dielectric loss and dielectric constant properties and used to describe the total dielectric properties of any material is mathematically expressed as:

$$\varepsilon^* = \varepsilon' \pm j\varepsilon'' \qquad (1)$$

where ε' is the real part of the complex dielectric constant that depicts the ability of the material to be polarized by an external electric field; and jε'' is the imaginary part of the complex dielectric constant representing the effective loss which quantifies the efficiency with which the microwave energy is converted to heat; j is an imaginary constant (Thostenson and Chou, 1999; Venkatesh and Raghavan, 2004).

Not all materials can be heated by microwave irradiation. Some materials, for example, all bulk metals reflect microwaves on the surface thus cannot be directly heated by microwaves. These materials are classified as conductors and are mostly used in the microwave applications to

67

make waveguides to transmit microwaves from the generator to the applicator. Conversely, materials which are transparent to microwaves are classified as insulators (e.g. teflon, glass, quartz, ceramic, etc.), while those which absorb the microwave energy are easily heated and classed as dielectrics (e.g. water, activated carbon, electrolytes, silicon carbide, magnetite, etc.). Generally, materials with a dissipation factor i.e. tangent loss (tan δ) > 0.5 are referred to as good microwave absorbers. Those with tan δ between 0.1 and 0.5 and those <0.1 are considered to be medium and low-microwave absorbers, respectively (Yin, 2012). A low loss material (e.g. insulator) can be blended with materials with a high loss factor (microwave facilitators) to facilitate an indirect microwave heating, where the microwave first heats the facilitator which then heats the low loss material (Haque, 1999; Menéndez et al., 2002). Figure 4-4 gives an illustration of the microwave interaction with the three classes of materials.

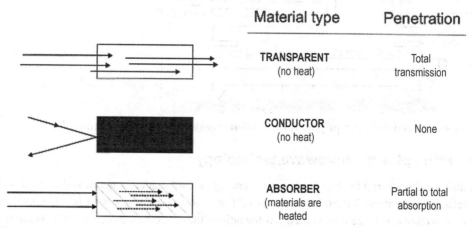

Figure 4-4. Interaction of microwave with materials (Haque, 1999)

4.4   Mechanisms of microwave heating

Heating of a material by microwave (i.e. dissipation of microwave energy to heat) is the result of two main mechanisms (Figure 4-5), namely rotation of dipolar species and polarization (migration) of ionic species (Haque, 1999). Dipole rotation, which is the primary heat-induction mechanism of microwave dielectric heating results from the interaction of the molecular dipoles with electromagnetic field. Dipolar molecules such as water have a random orientation. Whenever an electric field is applied, the molecules orient themselves according to the polarity of the field. However, in a microwave field, the polarity alternates very rapidly, e.g., at the 2,450 MHz microwave frequency, the polarity changes 2.45 billion cycles per second. This results in phase difference which causes molecular rotation as the field is always changing before the dipole re-orients to align itself. Such molecular rotation leads to friction with the surrounding medium, and heat is generated.

The ionic polarization occurs when an electrical field is applied to solutions containing ions. The ions of positive and negative charges move at an accelerated pace towards oppositely charged regions of the electrical field. This causes collisions between the ions and disruption of hydrogen bonds within water, both of which result in the generation of heat (Venkatesh and Raghavan, 2004).

68

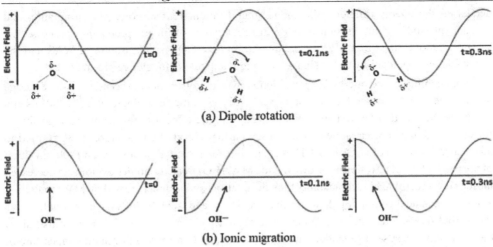

(a) Dipole rotation

(b) Ionic migration

Figure 4-5. Two main mechanisms of microwave heating: (a) dipole rotation; (b) ionic migration (Yin, 2012)

4.5 State of art of microwave application in waste treatment

Microwave (MW) irradiation technology has gained a wide range of use in various environmental applications such as pyrolysis, phase separation and extraction processes, soil remediation, remediation of hazardous and radioactive wastes, coal desulphurization, sewage sludge treatment, chemical catalysis and organic/inorganic syntheses (Remya and Lin, 2011). Particularly, there is emerging interest to use the technique for thermal treatment of waste sludge through sludge disintegration to enhance anaerobic digestion, to sanitise the sludge, to stabilise heavy metals, and to recover resources such as energy-rich biogas, bio-oil and nutrients (Tyagi and Lo, 2013).

In a study by Eskicioglu et al. (2006), it was found that thermal pre-treatment by MW irradiation at 96 °C was successful in disrupting the complex waste activated sludge (WAS) floc structure and releasing extra- and intra-cellular biopolymers, e.g. protein and sugars from activated sludge flocs into soluble phase along with solubilisation of particulate chemical oxygen demand (COD). Mesophilic anaerobic digestion of the solubilized sludge was also enhanced as indicated by a higher cumulative biogas production relative to control. Another study was conducted to assess the athermal (radiation) effects of microwaves on floc disintegration and consequently the anaerobic digestion of WAS relative to a conventionally heated (CH) control. Over a pre-treatment range of 50 - 96 °C, both MW and CH of WAS samples resulted in similar particulate COD and biopolymer solubilisation, and there was no apparent MW athermal effect on the COD solubilisation of WAS. However, the MW athermal effect enhanced the mesophilic anaerobic digestion of WAS as demonstrated by improved biogas production (based on biochemical methane potential (BMP) tests) for the MW samples over CH samples (Eskicioglu et al., 2007). Tang et al. (2010) investigated the efficiency of MW energy applied for pre-treatment of WAS. They observed that for a similar MW energy input, the solid solubilisation was higher in sludge samples with lower water content compared to those with higher water content. More energy was consumed to increase the temperature of the sludge with a higher water content, which decreased the energy efficiency in sludge solubilisation. Furthermore, MW irradiation and subsequent sludge solubilisation resulted in acceleration of biogas production process. The examples above show that MW treatment can improve COD solubilisation of WAS and the anaerobic biodegradability of the sludge, making it applicable for sludge pre-treatment.

The applicability of the microwave irradition in hybrid pretreatment processes of waste sludge has also been demonstrated at laboratory scale. For example, Eskicioglu et al. (2008) used a combined MW and H_2O_2 system to assess the synergetic effect of pre-treating sewage sludge for enhanced anaerobic treatment. They observed that elevated temperatures (>80 °C) led to more decomposition of H_2O_2 into ·OH radicals enhancing oxidation of COD and solubilisation of the particulate COD (>45μm). However, the MW/H_2O_2 treated samples exhibited a decreased mesophilic anaerobic biodegradability than the controls and samples treated by MW irradiation alone. In a study combining MW irradiation, H_2O_2 and acid hydrolysis in a single stage (MW/H_2O_2/H+ - (Advanced Oxidation Process) AOP), Chan et al. (2007) observed that all of the COD in the test sewage sludge was obtained in soluble form at a threshold temperature of 80 °C with H_2O_2 and H_2SO_4 dosages of 34-mL/L and 17 mL/L, respectively, indicating that all of the organic material was solubilised. Furthermore, Liao et al. (2007) observed an over 96% dissolution of TCOD after MW pre-treatment (20 °C/min) for 5 min with a 7-ml/L H_2O_2 dose (30% by wt).

A number of studies have also demonstrated the capability of the MW irradiation in sludge sanitization. Yu et al. (2010) studied the effect of a MW enhanced AOP treatment on pathogen destruction and regrowth, among other parameters, in activated sludge. The system successfully reduced faecal coliforms to below detection limit (1.0×10^3 CFU/L) immediately after treating the sludge at 70 °C with more than 0.04 % H_2O_2 (w/w), however, significant re-growth was observed for the treated samples after 72 h. Nevertheless, no regrowth was observed when the sludge was treated at 70 °C with 0.08 % H_2O_2 (w/w) or higher, suggesting a complete destruction of the faecal coliforms. Complete pathogen (e.g. faecal coliforms, total coliforms, *E. coli*, etc.) destruction was also reported in other studies in which various kinds of sewage sludge (e.g. primary, anaerobic digester and waste activated sludge) were heated by MW to temperatures above 65 °C (Hong et al., 2004; Hong et al., 2006; Tyagi and Lo, 2013). Pathogen destruction by MW was associated with both the athermal (electromagnetic radiation) and thermal (temperature) effects of the electromagnetic energy. Electromagnetic radiation causes the molecules of the irradiated material to orient themselves in the direction of the electric field, which may break hydrogen bonds leading to the denaturation and death of microbial cells (Banik et al., 2003; Tyagi and Lo, 2013). Conversely, the destruction by thermal effect is caused by the rapturing of microbial cells when water is rapidly heated to the boiling point by rotating dipole molecules under an oscillating electromagnetic field (Hong et al., 2004; Wojciechowska, 2005; Tang et al., 2010; Tyagi and Lo, 2013).

Successful reduction of sludge volume has also been achieved with MW irradiation. Menéndez et al., (2002) attempted a MW induced pyrolysis for the treatment of anaerobic sewage sludge by applying a microwave input power at 1 kW and 2.45 GHz frequency. They achieved a maximum temperature of 200 °C which was sufficient for sludge drying only, and attained a volume reduction over 80 %. The relatively low maximum pyrolytic temperature was associated with the poor absorbance of microwaves by the sludge. To enhance the absorbance, they mixed the wet sludge with char (a microwave facilitator) and achieved a temperature of 900 °C after 2 min of irradiation, which was sufficient for the pyrolysis.

4.6 Conclusions and recommendations

A review of the microwave heating shows its advantages over the conventional thermal heating processes. The potential application of the microwave technology in treating various sewage sludge streams has been successfully demonstrated. Most of the studies evaluated the potential of the microwave technology as a possible option for the sewage sludge pre-treatment prior to other processes such as the anaerobic digestion. The feasibility of the technology for sludge sanitization (pathogen destruction) and volume reduction have also been evaluated. Those evaluations have been successfully conducted at laboratory scale, but no evidence exists to demonstrate if they have been advanced and tested at pilot-scale. Furthermore, despite the successful results with the sewage sludge, the potential of microwave treatment of other common sludge types such as faecal and septic sludge, etc. has not been demonstrated. Yet based on its capability to rapidly and efficiently heat the matter, microwave can be a viable option for faecal and septic sludge treatment regarding volume reduction and pathogen inactivation. The technology would be more viable particularly in isolated situations with unusually high sludge generation rates such as the slums and emergency settlements. It is therefore desirable to evaluate the microwave treatment of faecal and septic sludge at laboratory scale, and subsequently scale up the technology and test with the common sludge types at pilot scale. If successful, the microwave irradiation may offer a rapid technological option for the sludge treatment in the areas where large quantities are generated.

References

Banik, S., Bandyopadhyay, S., Ganguly, S., 2003. Bioeffects of microwave—a brief review. Bioresour. Technol. 87, 155-159.

Bradshaw, S.M., Wyk, E.J.v., Swardt, J.B.d., 1998. Microwave heating principles and the application to the regeneration of granular activated carbon. J. S. Afr. Inst. Min. Metall. 98, 201-210.

Chan, W.I., Wong, W.T., Liao, P.H., Lo, K.V., 2007. Sewage sludge nutrient solubilization using a single-stage microwave treatment. Journal of environmental science and health. Part A, Toxic/hazardous substances & environmental engineering 42, 59-63.

Clark, D.E., Folz, D.C., West, J.K., 2000. Processing materials with microwave energy. Materials Science and Engineering: A 287, 153-158.

Eskicioglu, C., Kennedy, K.J., Droste, R.L., 2006. Characterization of soluble organic matter of waste activated sludge before and after thermal pretreatment. Water Res. 40, 3725-3736.

Eskicioglu, C., Prorot, A., Marin, J., Droste, R.L., Kennedy, K.J., 2008. Synergetic pretreatment of sewage sludge by microwave irradiation in presence of H2O2 for enhanced anaerobic digestion. Water Res. 42, 4674-4682.

Eskicioglu, C., Terzian, N., Kennedy, K.J., Droste, R.L., Hamoda, M., 2007. Athermal microwave effects for enhancing digestibility of waste activated sludge. Water Res. 41, 2457-2466.

Haque, K.E., 1999. Microwave energy for mineral treatment processes—a brief review. Int. J. Miner. Process. 57, 1-24.

Hong, S.M., Park, J.K., Lee, Y.O., 2004. Mechanisms of microwave irradiation involved in the destruction of fecal coliforms from biosolids. Water Res. 38, 1615-1625.

Hong, S.M., Park, J.K., Teeradej, N., Lee, Y.O., Cho, Y.K., Park, C.H., 2006. Pretreatment of Sludge with Microwaves for Pathogen Destruction and Improved Anaerobic Digestion Performance. Water Environ. Res. 78, 76-83.

Liao, P.H., Lo, K.V., Chan, W.I., Wong, W.T., 2007. Sludge reduction and volatile fatty acid recovery using microwave advanced oxidation process. Journal of environmental science and health. Part A, Toxic/hazardous substances & environmental engineering 42, 633-639.

Menéndez, J.A., Inguanzo, M., Pis, J.J., 2002. Microwave-induced pyrolysis of sewage sludge. Water Res. 36, 3261-3264.

NASA, 2016. Imagine the Universe. Goddard Space Flight Center, USA. http://imagine.gsfc.nasa.gov/science/toolbox/emspectrum1.html.

Remya, N., Lin, J.-G., 2011. Current status of microwave application in wastewater treatment— A review. Chem. Eng. J. 166, 797-813.

Tang, B., Yu, L., Huang, S., Luo, J., Zhuo, Y., 2010. Energy efficiency of pre-treating excess sewage sludge with microwave irradiation. Bioresour. Technol. 101, 5092-5097.

Thostenson, E.T., Chou, T.W., 1999. Microwave processing: fundamentals and applications. Compos. A: Appl. Sci. Manuf. 30, 1055-1071.

Tyagi, V.K., Lo, S.-L., 2013. Microwave irradiation: A sustainable way for sludge treatment and resource recovery. Renew. Sust. Energ. Rev. 18, 288-305.

Venkatesh, M.S., Raghavan, G.S.V., 2004. An Overview of Microwave Processing and Dielectric Properties of Agri-food Materials. Biosystems Engineering 88, 1-18.

Wojciechowska, E., 2005. Application of microwaves for sewage sludge conditioning. Water Res. 39, 4749-4754.

Yin, C., 2012. Microwave-assisted pyrolysis of biomass for liquid biofuels production. Bioresour. Technol. 120, 273-284.

Yu, Y., Chan, W.I., Liao, P.H., Lo, K.V., 2010. Disinfection and solubilization of sewage sludge using the microwave enhanced advanced oxidation process. J. Hazard. Mater. 181, 1143-1147.

Chapter 5
Evaluation of a microwave based reactor for the treatment of blackwater sludge

This chapter is based on:

Mawioo, P.M., Rweyemamu, A., Garcia, H.A., Hooijmans, C.M., Brdjanovic, D., 2016. Evaluation of a microwave based reactor for the treatment of blackwater sludge. Science of the Total Environment 548–549, 72-81.

Abstract

A laboratory-scale microwave (MW) unit was applied to treat fresh blackwater sludge, a similar substitute for faecal sludge (FS), produced at heavily used toilet facilities. The sludge was exposed to MW irradiation at different power levels and for various durations. Variables such as sludge volume and pathogen concentration reduction were observed. The results demonstrated that the MW is a rapid and efficient technology that can reduce the sludge volume by over 70% in these experimental conditions. The concentration of bacterial pathogenic indicator *E. coli* also decreased to below the analytical detection levels. Furthermore, the results indicated that the MW operational conditions including radiation power and contact time can be varied to achieve the desired sludge volume and pathogens reduction. MW technology can be further explored for the potential scaling-up as an option for rapid treatment of FS from intensively used sanitation facilities such as in emergency situations.

5.1   Introduction

Heavy usage of onsite sanitation facilities (i.e. 50 - 400 users per sanitation facility as observed in emergency settings) results in the rapid accumulation of fresh faecal sludge (FS) which should be frequently and safely disposed. Rapid accumulation rates result in the generation of large volumes of FS which can present a significant challenge for FS management especially during its transportation and disposal. The situation can be worsened by the generation of such huge amounts of FS in a locality with limited disposal possibilities. Particularly, if the availability of land is inadequate for local disposal, the FS may need to be hauled long distances to the final disposal site. In such situations, it might be economical to reduce the sludge volume in order to minimize the transport and ultimately the operation and maintenance costs of the sanitation system. Furthermore, FS contains various compounds of interest including high concentrations of organic and inorganic matter, and large amounts of pathogens. Pathogenic organisms found in FS include bacteria, viruses, protozoa, and helminths (Richard, 2001; Fidjeland et al., 2013). These organisms form a major concern to public health especially in the disposal and/or reuse of the sludge. The pathogenic organisms should thus be reduced to safe levels (e.g. *E. coli* $\leq 1.0 \times 10^3$ CFU/gTS (WHO, 2006)) in order to minimize public health risk posed by the possible outbreaks of faecal sludge related epidemics. The presence of organic matter in the FS is yet another important aspect in the FS management as it can lead to offensive odors and attract vector organisms (such as houseflies and mosquitoes) that can spread diseases. Therefore, organic stabilization of the FS is desirable to ensure safe waste disposal practice. The aspects highlighted above form a major challenge in FS management but can be addressed by the use of appropriate FS treatment technologies.

A number of FS conventional treatment options are available, including the conventional drying (e.g. in sludge drying beds), composting, co-composting with organic solid waste, anaerobic co-digestion with organic solid waste (producing biogas), and co-treatment in wastewater treatment plants (WWTP) (Ingallinella et al., 2002; Ronteltap et al., 2014). These technologies have been tested and applied in regular sanitation contexts. They have associated benefits, but also limitations that make their application less suitable in some specific contexts such as the emergency situations. For instance, the composting technology produces a hygienically safe product rich in humus; however, it requires much space, long treatment duration, and may pose environmental pollution and public health risks in low-lying areas in the case of flooding (Katukiza et al., 2012). Anaerobic co-digestion with organic solid waste offers the benefit of both increasing biogas production, as well as using the final (end) product as a fertilizer. However, this option has the limitation of involving a relatively slow digestion process. In addition, a post treatment stage is required for the further removal of pathogens (Katukiza et al., 2012). Co-treatment in WWTPs is a possible option for FS treatment, but the probable heavy hydraulic, organic, and solids loads may limit the application of this alternative (Lopez-Vazquez et al., 2014), especially, if the co-treatment of sludge was not considered in the original design.

Generally, the major limitations of conventional FS treatment technologies highlighted above include their relatively slow treatment processes and large space requirements making them less feasible in scenarios with high FS generation rates and limited land space. Such scenarios are commonly faced when dealing with heavily used onsite facilities such as during an emergency situation. Inappropriate FS treatment in emergencies has, in some cases, resulted in the adoption of poor disposal methods, especially in less developed countries where common practice is to use pit latrines and septic tanks. These facilities fill up rapidly when intensively used, in which case the adoption of poor disposal methods is likely to occur. A recent example is the open dumping of raw FS that was reported in Haiti after the earthquake in 2010 (Oxfam, 2011). Such poor FS management practices pose great dangers to affected people whose public health is already jeopardized by the poor living conditions in the disaster environment. For instance, the rapid spread of a cholera epidemic after the Haiti earthquake in 2010 which claimed approximately 500,000 lives was associated with inadequate sanitation provision (Tappero and Tauxe, 2011). Furthermore, sanitation related outbreaks of diarrheal diseases were reported after the earthquake in Pakistan in 2005, the tsunami in Indonesia in 2004, the floods in Bangladesh in 2004, and the floods in Mozambique in 2000 (Watson et al., 2007).

These challenges for FS treatment in areas with high generation rate and limited land space demonstrate the need to explore more options which are fast and efficient. MW technology may present an appropriate alternative for future applications in FS treatment in such areas. The MW irradiation has characteristics such as instant and accurate control of the power input as well as providing fast and uniform heating throughout the target material (Haque, 1999). With its unique nature in rapid heating, the MW technology application looks very promising for FS treatment in the situations requiring rapid treatment options. Furthermore, MW based applications can potentially provide compact and easily portable as well as fast and effective FS treatment package units with reduced footprints. The MW technology uses microwave energy which is a non-ionizing electromagnetic radiation with wavelengths between 1mm and 1m and frequencies between 300MHz and 300GHz (Haque, 1999; Tang et al., 2010; Remya and Lin, 2011). The technology has been widely used in communication, industrial, scientific, medical and instrumentation applications (Haque, 1999). Most of the above applications utilize the technology for heating where the microwaves cause molecular motion in the target material by inducing the migration of ionic species and/or the rotation of dipolar species (Haque, 1999; Thostenson and Chou, 1999). The heating by microwaves depends on the dissipation factor, which is the ratio of the dielectric loss factor to the dielectric constant of the target material. The dielectric loss factor depicts the amount of MW energy lost in the material by dissipation in form of heat while the dielectric constant depicts the ability of material to delay or retard microwave energy as it passes through. Therefore, materials that are easily heated by MW energy have a high dielectric loss factor (Haque, 1999; Thostenson and Chou, 1999).

The MW technology was applied in the treatment of some common wastes such as the sewage sludge that contains dipolar molecules (e.g. water and organic complexes) with high loss dielectric properties that enable selective and concentrated heating by microwaves (Yu et al., 2010). A number of studies that were conducted using the various types of sewage sludge demonstrated the success of MW treatment in many aspects including pathogen reduction (Tyagi and Lo, 2013). For instance, complete pathogen destruction was reported when primary, anaerobic digester and waste activated sludges were heated by MW to temperatures above 65°C (Hong et al., 2004; Hong et al., 2006; Tyagi and Lo, 2013). In addition, over 80% volume reduction was reported when anaerobic sewage sludge was exposed to MW irradiation (Menéndez et al., 2002). The pathogen destruction by MW was associated with both the non-thermal (electromagnetic radiation) and thermal (temperature) effects of the electromagnetic energy. Electromagnetic radiation causes the molecules of the irradiated material to orient themselves in the direction of the electric field. This may result in hydrogen bonds breaking leading to the denaturation and death of microbial cells (Banik et al., 2003; Tyagi and Lo, 2013). On the other hand, the destruction by thermal effect is caused by the rotation of dipole molecules under an oscillating electromagnetic field which results in rapid heating of water to boiling point. The cells of microorganisms are then ruptured and the bound water is released (Hong et al., 2004; Wojciechowska, 2005; Tang et al., 2010; Tyagi and Lo, 2013).

Such successful applications in the sewage sludge treatment demonstrate the potential of the MW technology in treating FS. Although different in aspects such as concentration of organics, pathogen, TS, among others, the fresh FS has relatively similar dielectric properties to that of sewage sludge. For instance, like the sewage sludge, FS contains dipolar (e.g. water and organic complexes) molecules which are important in the MW heating. The dipolar characteristic of the fresh FS thus provides an opportunity for its treatment by the MW technology.

Despite the successful evaluations of the MW treatment using various kinds of sewage sludge, no studies have yet been reported with respect to FS context. It is thus desired to evaluate MW application on FS, since FS is more concentrated, and comparatively has more pathogens, and organics, among others. Therefore, the objective of this study is to investigate the potential of a microwave (MW) based technology for treating fresh blackwater FS extracted from a highly concentrated domestic blackwater stream. The study focused on three aspects of treatment namely the volume reduction (drying), sanitization (bacterial pathogen reduction), and organic stabilization (organic matter reduction) in the sludge. The weight was used to estimate the volume reduction while the removal efficiency of *E. coli* was used as an indicator for the reduction of bacterial pathogens. The volatile and total solids ratio (VS/TS) was used as an indicator for the organic stabilization of the sludge. This is arguably the first study to evaluate the use of MW technology for fresh FS treatment. If successful, this technology may provide a solution to the complex task of FS treatment; particularly, when dealing with heavily used onsite sanitation facilities such as in emergency settings.

5.2 Materials and methods

5.2.1 Research design

This study was performed using blackwater FS which was extracted by centrifugation from autoclaved and non-autoclaved highly concentrated fresh blackwater stream obtained from the DESAR (Decentralized Sanitation and Reuse) demonstration site in Sneek (Friesland, NL). The 20 g sample was autoclaved to remove all the existing organisms in the sludge, and then a known concentration of harmless *E.coli* was introduced and its response to MW treatment was closely observed. Then the study was advanced with the non-autoclaved 100 g sample in which the *E.coli* naturally occurring in the sludge was monitored. Being a type of FS (Strande, 2014), blackwater could be directly used in this study. However, the aim was to evaluate the MW treatment with a more concentrated FS stream, similar to what is generated in emergency situations where non-flush toilet facilities are commonly applied. And since this kind of FS could not be obtained in The Netherlands, as the country is mostly sewered, the centrifugation of blackwater to obtain a more concentrated FS was considered. Two sludge fractions (i.e. 20 g with the autoclaved sludge and 100 g with the non-autoclaved sludge) were treated in a domestic MW oven for various durations and input MW power levels. Changes in temperature, weight reduction (volume indicator), *E. coli* (pathogen indicator) and VS/TS ratio (organic matter indicator) were measured in the samples treated by MW. The experiments using the 20 g sludge were conducted in duplicate while those using 100 g sludge were conducted in triplicate. The effectiveness of the MW treatment was evaluated based on changes in the measured parameters before and after exposure to MW.

5.2.2 Microwave apparatus

A domestic MW oven, Samsung, MX245 (Samsung Electronics Benelux B.V., the Netherlands) was used in this study Figure 5-1.

Figure 5-1. The domestic microwave unit (Photo: P. Mawioo)

The unit operates at a frequency of 2450 MHz with a power output ranging from 0 to 1,550 W with 10 % incremental steps.

5.2.3 Sludge samples

The blackwater samples were drawn from a buffer tank receiving blackwater collected from vacuum toilets flushed with approximately 0.5 L of water per use. The samples were then transported to the research laboratory and stored at 4°C prior to the experiments within 48 hours. The proximate characteristics of the fresh blackwater are presented in Table 5-1. The blackwater was concentrated by centrifugation to attain a blackwater FS with proximate characteristics as presented Table 5-1, which are common for FS from non-flush toilet facilities (Fidjeland et al., 2013).

Table 5-1. Physico-chemical characteristics of fresh blackwater and the blackwater FS

Parameter	Blackwater Value	Blackwater FS Value
Water content (%)	98.6	88
Total Solids, %	1.4	12
Volatile Solids, %	1.3	10.7
pH	6.9 - 7	6.7
Total COD, TCOD (mg/g TS)	1643	1344
E. coli (CFU/g TS)	2.3×10^6	4.0×10^8

5.2.4 Experimental procedures

Sample preparation

The characteristics of the fresh blackwater FS sample that was applied in this study are presented in Table 1 above. The blackwater (Figure 5-2a) used to prepare the 20 g samples was autoclaved at 121°C for one hour in a standard autoclave (Tuttnauer, model 3870 ELV, Tuttnauer Europe B.V., Breda, The Netherlands) to destroy all existing organisms. The sample was then concentrated using a bench-top centrifuge (ROTINA 420, Hettich, Germany) operated at a relative centrifugal force (RCF) of 1,851 for 30 minutes. Following this, the supernatant was discarded and the resulting sludge cake (see Figure 5-2b) (i.e. sludge cake, TS = 12 %) was spiked with a harmless *E. coli* (type ATCC25922) to the final concentration of 10^8 CFU/g TS. The test samples for the MW treatment were prepared in duplicates by placing 20 g samples (height approximately one centimeter, and surface area approximately 33.2 cm^2) in a 250 mL glass beaker. Similarly, the 100 g samples were prepared in the same procedure explained above but with non-autoclaved fresh blackwater. The resulting sludge cake (for the non-autoclaved 100 g samples) was not spiked as it contained the *E. coli* naturally occurring in human excreta (10^8 CFU/g TS, TS =12 %). The test samples for the MW treatment were prepared in triplicates by placing 100 g samples (height approximately one centimeter, and surface area approximately 78.5 cm^2) in a 1000 mL glass beaker. In both cases, *E. coli* was used as an indicator to evaluate the MW capability to destruct the pathogenic bacteria in the sludge.

Figure 5-2. (a) Blackwater and (b) blackwater FS (sludge cake) obtained after centrifugation (Photo: P. Mawioo)

Microwave treatment

Both the 20 g (autoclaved) and 100 g (non-autoclaved) sludge samples were treated using the MW apparatus described in the Section 5.2.2 above. The sample contained in the glass beaker was placed in the MW cavity and exposed to the MW irradiation at 465, 1,085, and 1,550 W for varied time durations. The 20 g samples were exposed for 5, 10, 20, 30, 60, 120, 180, and

240 seconds (i.e. 0.08, 0.17, 0.33, 0.5, 1, 2, 3, and 4 minutes, respectively), while the 100 g samples were exposed for 1, 3, 5, 7, and 10 minutes. After the MW treatment, the sample was removed from the MW cavity and its temperature was immediately measured before covering with sanitized aluminium foil. The microwaved samples (Figure 5-3) were cooled to room temperature and analyzed for their characteristics as described in the following section.

Figure 5-3. The MW treated blackwater FS test samples

5.2.5 Analytical procedures

Sludge samples with and without MW treatment were measured for physical, chemical and microbial parameters including the total COD (TCOD), temperature, weight/volume reduction, TS, VS, and *E. coli*.

COD measurement

Samples for COD measurement in the blackwater (prior to centrifuging) and the blackwater FS (prior to MW treatment) were prepared by diluting a known amount in demineralized water. The COD measurement was then done according to the closed reflux method (SM 5220 C) as outlined in the Standard Methods for the Examination of Water and Wastewater (APHA, 2012). The values were expressed in mg COD per g TS (mg COD/g TS) as shown in Table 1 (Section 2.3).

Temperature measurement

The initial sample temperature was measured just before MW irradiation using an infrared thermometer (Fluke 62 MAX, Fluke Corporation, U.S.A). Following each treatment, a sample was taken from the MW cavity and the final temperature was immediately measured. While measuring the final temperature, samples were mixed by shaking to avoid taking only temperature at the surface of the heated sludge. However, for those samples that were too dry to mix by shaking (TS $\geq$17%), the temperature was measured on the surface.

Weight measurement

Samples weight measurements were done using a bench-top weighing balance (Sartorius H160, Sartorius AG, Germany). The initial weight was measured as the samples were transferred into the heating beakers, while the final weight was measured once the samples were cooled to room temperature after the MW treatment. The volume reduction was then determined from the difference between the initial and the final sample weight. Based on the maximum temperature attained during MW treatment (i.e. $\leq 127^\circ C$), the weight reduction could only be attributed to the water evaporating from the heated sludge. Thus, considering the density of water, the weight reduction was deemed to be equivalent to the sludge volume reduction.

Microbial measurement

The detection of *E. coli* was done using the surface plate technique with chromocult coliform agar (Chromocult; Merck, Darmstadt, Germany) (Byamukama et al., 2000). Portions of one gram from each MW treated sample and the control (untreated sludge) were transferred to sanitized plastic containers (20 mL), then mixed with nine ml of buffered peptone water and thoroughly homogenized. A potter tube (Potter-Elvehjem PTFE pestle and glass tube, Sigma-Aldrich Co. LLC, USA) was used to grind those samples that were too dry to directly dissolve in the buffered peptone water. The homogenized samples were serially diluted (10^{-1} to 10^{-5}) with the buffered peptone water. Partitions of 100 µL of the respective sample dilutions were applied to the chromocult agar plates in duplicate for each dilution step. All plates were then incubated at $37^\circ C$ for 24 hours, after which the average numbers of colonies in plates were counted. The counting was visually facilitated by a colony counter (IUL magnifying glass colony counter, IUL, S.A., Barcelona, Spain). Dark blue to violet colonies were classified as *E. coli* (Byamukama et al., 2000; Sangadkit et al., 2012). The average number of colonies was used to calculate the viable-cell concentrations in the solid samples, expressed in CFU/ g TS of the test sample.

TS and VS measurement

For each treated sample the TS and VS values were measured according to the gravimetric method (SM-2540D and SM-2540E) by drying a known sample weight in an oven at $105^\circ C$ for two hours (for TS) and subsequently in a muffle furnace at $550^\circ C$ for two hours (for TVS) as outlined in the Standard Methods for the Examination of Water and Wastewater (APHA, 2012). The TS and VS results were then used to evaluate sludge stability based on the organic matter reduction.

5.3 Results

5.3.1 Temperature evolution

Figures 5-3a and 5-3b show the temperature profiles during MW exposure of the 20 g and 100 g sludge sample, respectively. As expected, the temperature increment rate increased as the MW power input rose, with 1,550 W inducing the highest rate. As shown in Figures 5-3a and 5-3b, three phases were observed in the temperature rise during the MW heating. The first phase shows a rapid increase in sludge temperature, the second phase shows a fairly constant and minimal temperature rise, and the third phase depicted in again a rapid temperature increase. However, in this case the third phase was only achieved when the sludge was heated at 1,550 W at contact time above three and seven minutes for the 20 g and 100 g samples, respectively.

This implies that for the 465 W and 1,085 W, contact times longer than four minutes for 20 g and 10 minutes for 100 g are required to attain the third temperature phase.

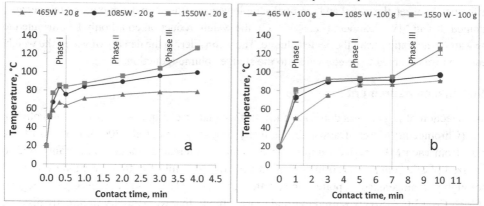

Figure 5-4. Effect of exposure to MW irradiation on temperature in the a) 20 g sludge sample and b) 100 g sludge sample

5.3.2 Volume reduction and energy requirements

Figures 5-4a and 5-4b show the respective weight/volume reduction when the 20 g and 100 g sludge samples were exposed to different MW power input levels and contact times. The weight/volume reduction is mainly associated with the temperature changes and the resulting moisture loss in the irradiated sludge. The weight/volume reduction process seems to occur in three stages which closely follow the trend of the three temperature phases described in Section 5.3.1 above. These drying stages were identified as the preliminary, essential (major), and final drying phases (Flaga, 2005). In the 20 g sample size (as illustrated in 5-4a), the preliminary drying phase occurred within the first 10 seconds for the 1,085 W and 1,550 W power levels corresponding to 4 and 5.5% volume reduction, respectively. However, this preliminary drying phase lasted longer, until 30 seconds at 465 W with approximately 5% weight/volume reduction. Immediately after the preliminary phase the treatment entered the essential (major) drying phase depicted by high and relatively constant moisture evaporation rates. The duration for the essential drying phase varied with the power input level. For instance, the essential drying phase lasted until approximately 2.5 and two minutes for the 1,085 and 1,550 W, respectively, while this phase was not conclusively achieved at the 465 W in the range of the contact times evaluated in these experiments. The final drying phase was achieved only with the 1,085 and 1,550 W power inputs and depicted by the lowest observed sample weight/volume reduction. Similarly, the weight/volume reduction profiles for the 100 g samples were observed (as illustrated in Figure 5-4b) and the trends were comparable to that reported on the 20 g sample. For instance, the preliminary drying phase lasted for approximately one minute (60 seconds) in the 1,085 W and 1,550 W power input levels with approximate weight/volume reductions of 2.7 and 4.5%, respectively. Furthermore, at 465 W power input level, the preliminary drying phase extended to approximately three minutes (180 seconds) with 2.8% weight/volume reduction. Moreover, in a trend similar to that observed in the 20 g sample, the duration of the entire essential (major) drying phase in the 100 g sample varied with the power input. This phase lasted until seven minutes (420 seconds) when the sample was irradiated at

82

1,550 W, but it was not conclusively achieved at 465 and 1,085 W in the range of contact times evaluated in these experiments. High weight/volume reductions (over 80%) were achieved for both sample sizes in which a big fraction of the weight/volume reduction is associated with the major (essential) drying phase.

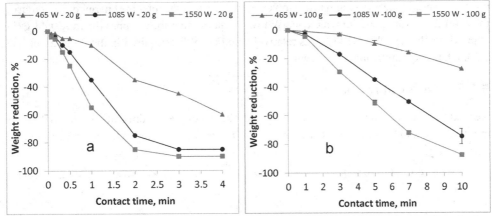

Figure 5-5. Effect of exposure to MW irradiation on sludge weight in the a) 20 g sludge sample and b) 100 g sludge sample

Furthermore, the energy consumption profiles during the MW heating were observed for both the 20 g and 100 g samples and they are presented in Figures 5-5a and 5-5b, respectively. The trends depicted in the energy consumption profiles correspond with the preliminary, essential and final drying phases that were previously discussed in this section.

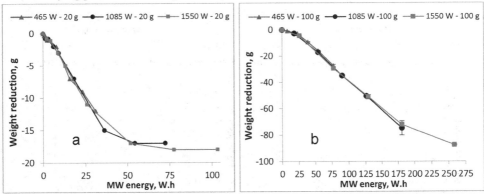

Figure 5-6. Weight reduction and MW energy demand in the a) 20 g sludge sample and b) 100 g sludge sample

A linear regression was performed on each of the three drying phases to reveal the corresponding specific energy demand rates in both the 20 g and 100 g sample. It was observed that for the 20 g sample, 1 watt-hour (Wh) was required to remove approximately 0.25 g (i.e. approximately 4 kWh per kg) during the preliminary drying. In the essential drying phase, where the lowest energy demand was observed, one Wh was required to achieve approximately 0.4 g reduction (i.e. approximately 2.5 kWh per kg). The final drying phase marked by relatively high energy input with low corresponding weight/volume reduction was achieved only at the 1,550 W power input level, where one Wh was required to remove approximately 0.21 g (i.e.

approximately 4.8 kWh per kg). It was also observed that the specific energy demand rates in the 100 g sample showed similar trends to that observed in the 20 g sample. In this case, one Wh was required to remove approximately 0.12 g (i.e. approximately 8.3 kWh per kg), which was almost twice the energy demand compared to that in the 20 g sample (i.e. 4 kWh per kg), during the preliminary drying phase. However, in the 100 g sample, the energy demand during the essential drying phase in which one Wh was required to remove approximately 0.44 g (i.e. approximately 2.3 kWh per kg), was relatively similar to that required in the 20 g sample (i.e. 2.5 kWh per kg) under the same drying phase

5.3.3 Bacterial reduction

Reduction results of the *E. coli* in the 20 g and 100 g sample obtained at various MW power input levels and contact times are shown in Figures 5-6a1 and a2, and 5-6b1 and b2, respectively.

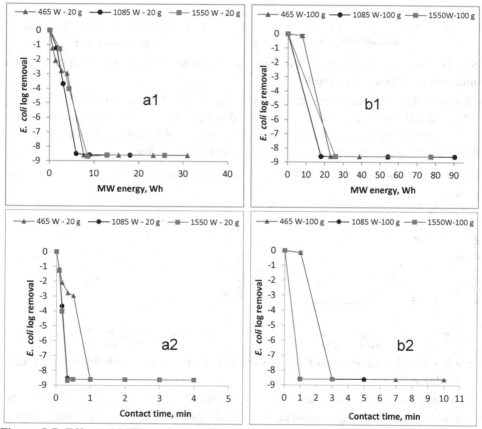

Figure 5-7. Effect of MW energy on *E. coli* reduction in a1) 20 g sludge sample and b1) 100 g sludge sample, and *E. coli* reduction as a function of time in a2) 20 g sample and b2) 100 g sample. The zero *E.coli* log removal corresponds to an initial concentration of 4.0 x 10^8 CFU/g TS).

Furthermore, the influence of temperature on the *E. coli* reduction over contact time in the 20 g and 100 g sample is illustrated in Figures 5-7a and 5-7b, respectively.

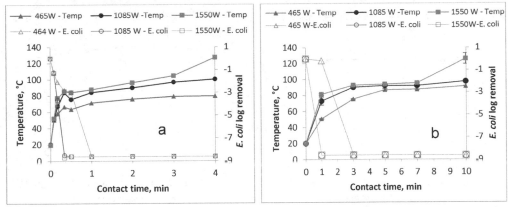

Figure 5-8. Influence of temperature on *E. coli* reduction in a) 20 g sludge sample and b) 100 g sample. The zero *E.coli* log removal corresponds to an initial concentration of 4.0 x 10^8 CFU/g TS).

The results show that the increases in MW power input and/or contact time (see Figures 5-6a1, 5-6a2 and 5-6b1, 5-6b2) and the sludge temperature (see Figures 5-7a and 5-7b) led to increased *E. coli* reduction. For instance, in the 20 g sample, an *E. coli* reduction of approximately three log removal value (LRV) was achieved when the sludge was MW treated at 465 W over a 0.5 minute (30 seconds) contact time (i.e. MW energy = 4 Wh, temp. = 63°C). However, reduction below detection limit (i.e. < 1,000 CFU/g TS) was achieved with one minute contact time (i.e. MW energy = 8 Wh, temp. = 71°C). The exposure time for reduction to below the detection limit was shorter at 0.5 minute (30 seconds) (i.e. MW energy = 6 Wh, temp. > 70°C) when the sludge was irradiated at higher MW power levels equal to or higher than 1,085 W. Similarly, while the MW treatment at 465 W for one minute (MW energy = 8 Wh, temp. = 51°C) depicted the lowest *E. coli* reduction (approximately 0.2 LRV) in the 100 g sample, a reduction to below the detection limit was realized at the same power level when the contact time was increased to three minutes (MW energy = 23 Wh, temp. = 75°C). However, *E. coli* reduction below detection limit could still be achieved at one minute when higher MW power input levels (e.g. 1,085 W, MW energy = 18 Wh, temp. = 73°C and 1,550 W, MW energy = 26 Wh, temp. = 81°C) were used.

5.3.4 Effect of microwave irradiation on organic stabilization of sludge

The VS/TS ratio in the treated samples was used as an index to determine the organic stability of the blackwater FS. Figures 5-8a and 5-8b show the variation in the VS/TS ratio as a function of MW power and contact time in the 20 g and the 100 g samples, respectively. As shown in Figures 5-8a and 5-8b, the MW treatment was not successful in organic matter reduction as there was no significant change in the VS/TS ratio in the treated sludge. For instance, the respective initial VS/TS ratios in the 20 g and 100 g untreated samples was approximately 80% and 89%, while the final VS/TS ratio attained at all power input levels and contact times evaluated were in the same range.

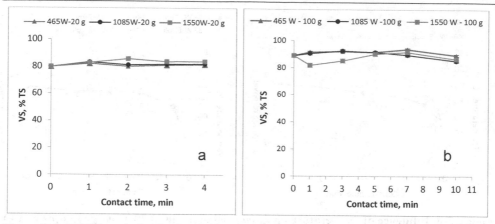

Figure 5-9. VS/TS variation in the a) 20 g and b) 100 g MW treated sludge sample

5.4 Discussion

5.4.1 Temperature evolution

Results from Figures 1a and 1b show that MW treatment was very effective and fast in raising the temperature in the blackwater FS. The temperature evolution in both the 20 g and 100 g sample was depicted in three phases classified as the preliminary, essential (major), and final drying phases. These temperature evolution phases conform to those reported in other drying methods such as convection and conduction (Flaga, 2005; Bennamoun et al., 2013). However, Bennamoun et al., (2013) gave a different terminology for the three phases, namely the adaptation phase, constant drying rate phase, and falling drying rate phase. The trends observed in this study also agree with those from previous studies involving MW heating of different kinds of sewage sludge (Menéndez et al., 2002; Hong et al., 2004; Yu et al., 2010; Lin et al., 2012). Furthermore, the results show that the heating rate increases with the MW power input increment which is similar to the trend reported in other studies (Eskicioglu et al., 2007; Lin et al., 2012) when sewage sludge was heated at varied MW power input levels. This can be explained by the resulting MW energy which is a function of the power input.

The rapid rise in temperature observed during the first (preliminary) drying phase can be attributed to the interaction between the microwaves (i.e. high frequency electromagnetic radiation) with the dipolar molecules of high loss dielectric properties (e.g. water, proteins, etc.) that are initially present at high concentrations in the wet blackwater FS. This interaction causes the molecular rotation resulting in the rapid heating of the sludge (Yu et al., 2010). As expected, the temperature increased more rapidly in the 20 g samples than in the 100 g samples. This can mainly be attributed to the different amounts of water content in the different sample sizes. Water has a high thermal capacity, thus by virtue of its higher water content, the bigger sample size (100 g) has a higher capacity to absorb a bigger fraction of the initial MW energy. This occurs with a relatively smaller temperature increase rate than in the smaller sample size (20 g). Similar observations were made by Tang et al., (2010) when excess sewage sludge was heated with different water contents. In the second (essential/major) drying phase, the sludge may have reached boiling point as it was characterized by a fairly constant and minimal rise in temperature. At this stage the unbound water evaporates from or near the surface of the sludge particles at constant rate. The sludge particles are covered by water on the surface that constantly evaporates as it is replaced by water from inside the particles (Flaga, 2005). As the unbound water is depleted, the heating entered the third (final) drying phase in which the sludge temperature begun to rise rapidly. According to Flaga, (2005), in this phase, water on the surface of the particles evaporates faster than it is replaced by water from inside the particle. In this study, the final drying phase was only realized when sludge was heated at 1,550 W at contact time above three and seven minutes for the 20 g, and 100 g samples, respectively. That is, the final drying phase was not attained at 465 W and 1,085 W demonstrating that longer contact time is required once the MW power input level is reduced. This is also true for the bigger samples that need more contact time to achieve similar results as in the smaller samples. If heating is continued in the final drying phase, it appears that the temperature rise reaches a certain maximum after which there is hardly any significant temperature increase. For instance, Menéndez et al., (2002) reported a maximum temperature of approximately 200°C with MW heating of sewage sludge. However, the maximum attainable temperature varies from material to material as was demonstrated when a MW receptor material was mixed with sewage sludge to raise the maximum temperature to over 900°C (Menéndez et al., 2002; Menéndez et al., 2005).

The results of this study show that temperature evolution for the MW heated blackwater FS follows similar trend as observed in sewage sludge when heated either by MW or conventional thermal technologies. Furthermore, as confirmed in several studies that compared MW to the conventional heating (e.g. water bath (Hong et al., 2004) and electric furnace (Menéndez et al., 2002)) in sewage sludge, the MW is superior in terms of the temperature evolution rate. Based on this observation and the fact that temperature has been reported to play an important role in the MW treatment of sludge (Banik et al., 2003; Hong et al., 2004; Shamis et al., 2008), a proper MW based full scale designated reactor will require shorter contact time than a conventional thermal reactor to achieve the same level of treatment. This in turn implies savings on time and ultimately the reactor space requirements when MW irradiation is applied.

5.4.2 Volume reduction and energy requirements

As shown in Figures 5-4a and 5-4b, the MW treatment was successful in achieving over 70% weight/volume reduction in the blackwater FS within given exposure boundaries. This conforms to the results obtained by Menéndez et al. (2002) who reported 80% volume reduction in anaerobic sewage sludge. The variations in weight/volume reduction in both the 20 g and 100 g sample closely followed the temperature profiles and the drying phases discussed in Section 5.4.1.

Generally, the results show low weight/volume reduction in both sample sizes during the preliminary drying phase. The low weight/volume reduction can be attributed to the minimal moisture loss since the MW energy initially supplied is largely utilized for sludge temperature elevation to the boiling point and not for the water evaporation.

However, a high but relatively constant weight/volume reduction rate was observed in both samples during the essential (major) drying phase that occurred immediately after the preliminary drying phase. The duration of the entire essential drying phase within the same sample decreased with the increase in the MW power input levels. This can be attributed to the rise in the MW energy resulting from the increased power input. Moreover, the length of the essential drying phase varied with the sample size and was shorter in the smaller sample. For instance, when the 20 g sample was irradiated at 1,550 W, the essential drying phase lasted until three minutes while it lasted until seven minutes at the same irradiation energy for the 100 g sample. This can be explained by the total amount of water in the blackwater FS to be removed during the essential drying phase which is smaller in the small sample. Generally, the high weight/volume reduction in the essential drying phase is achieved mainly due to the removal of the free (unbound) water which requires less energy.

The third (final) drying phase was depicted by the lowest observed moisture loss in both the 20 g and 100 g sample. The low weight/volume reduction manifested in this stage is attributable to the fact that much of the water is previously evaporated in the essential drying phase and weight/volume reduction is only possible by evaporating the bound water that requires much more energy. As reported in Flaga, (2005), the speed of drying at this stage decreases until it reaches a balanced hydration which is dependent on the heating temperature and the air humidity.

Similar trends in the drying phases discussed above were observed when dewatered sediment sludge was subjected to MW drying (Gan, 2000).

The energy consumption rate profiles shown in Figures 5-5a and 5-5b were closely associated with the weight/volume reduction and drying phases discussed above. Differences in the energy consumption between the two sample sizes were observed, especially during the preliminary and final drying phase. The discrepancy in the energy consumption during the preliminary phase (e.g. approximately; 8.3 kWh per kg and 4 kWh per kg for 100 g and 20 g, respectively) is attributable to the fact that much of the energy at this phase is not utilized in the actual drying but in raising the sludge temperature towards boiling point. Consequently, higher energy demand is possible in the 100 g sample which has higher initial water and TS content than the 20 g sample. Furthermore, the final drying phase (only achieved here with the 1,550 W) was also marked by relatively high energy demand attributable to the removal of the bound water which is more difficult to evaporate.

The lowest energy demand was observed during the essential (major) drying phase with no significant difference in the energy consumption rates between the two sample sizes. For instance, the respective energy consumption for the 100 g and 20 g sample was approximately 2.3 kWh per kg and approximately 2.5 kWh per kg. This shows that the energy supplied at this phase is mainly used in the evaporation of water, hence the similarity in the energy consumption rates between the samples. The low energy demand at this phase can be attributed to the presence of the free (unbound) water which is relatively easy to remove once the sludge is heated. Since water forms a bigger proportion of FS (approximately 88% in this case), it is evident that the essential drying phase is the most crucial phase for weight/volume reduction during the MW heating of the sludge. This implies that irradiation should be stopped at some point within the essential drying phase if the sludge is MW heated for the purpose of drying only. This is reasonable because there is no significant weight/volume reduction expected beyond this stage as manifested in the high specific energy demand in the ensuing final drying phase. The optimal point where to stop the heating within the essential drying phase will depend on the ultimate weight/volume reduction or moisture level desired and the associated energy costs. As discussed above, during the essential (major) drying phase, the specific energy requirements were relatively similar both between the two sample sizes and among the MW power levels evaluated in this study. The only difference was the contact time that was required to build up energy to the level corresponding to a certain amount of weight/volume reduction. This duration can thus be taken as the retention time of the MW reactor, a key factor in the full scale design that will affect its volume. This implies that when designing a MW reactor for applications in areas with high sludge generation rates but limited land space (e.g. the emergency camps), high MW power input with short retention time will be desirable to achieve a small reactor volume (i.e. small footprint). However, the retention time should be carefully chosen to ensure that other treatment objectives (e.g. pathogen reduction) are achieved.

Despite the fairly low energy demand, it is notable that the specific energy consumptions attained in the essential drying phase here are higher than in other drying methods, especially the convective and conductive industrial driers which according to Bennamoun et al., (2013) vary between 0.7-1.4 and 0.8-1.0 kWh, respectively, per kg of evaporated water. The cause of this disparity may be twofold. Firstly, the specific energy consumptions as reported by Bennamoun et al., (2013) are based on sewage sludge whose properties may differ from that of the blackwater FS, especially the viscosity which may hinder the microwaves penetration capacity (Hong et al., 2004). Secondly, the lack (in the unit used in the this study) of some important design aspects that are found in the industrial driers such as customized ventilation for moisture extraction may also be a major contribution to the disparities. Thus the energy demand here, especially during the essential drying phase may possibly be reduced by customizing the unit's design for maximum vapour extraction during the treatment. Furthermore, the results obtained here are only preliminary, and thus more research is needed to optimize the system, increase efficiency and reduce energy input. More tests with different kinds of sludge are also needed.

5.4.3 Bacterial reduction

The results shown in Figures 5-6 and 5-7 demonstrate that MW irradiation was very rapid and effective in the reduction of *E. coli* in blackwater FS. This observation is consistent with those from previous studies (Border and Rice-Spearman, 1999; Lamb and Siores, 2010), although these studies used different sample material. The destruction of the *E. coli* during the MW irradiation treatment can be attributed to both the non-thermal (electromagnetic radiation) and the thermal (temperature) effects. The electromagnetic radiation effect has been identified a factor in pathogen destruction in MW treatment (Banik et al., 2003; Hong et al., 2004; Shamis et al., 2008), but the temperature effect is considered the main mechanism for which 70°C is identified as the minimum temperature essential for complete bacterial reduction (Hong et al., 2004; Tyagi and Lo, 2012; Valero et al., 2014). As observed here, for any given power level, the contact time necessary for complete bacterial reduction will vary with the amount of material treated. This is clearly illustrated in the two sample sizes evaluated. For instance, *E. coli* was detected when the 100 g sample was irradiated at 465 W for one minute (MW energy = 7.75 Wh, temp. = 51°C) while none was detected in the 20 g sample treated under similar operational conditions. The disparity in reduction between these two sets of samples can be attributed to the observed differences in their final temperature (i.e. 71°C in 20 g sample and 51°C in 100 g sample (Figure 5-7a and 5-7b)). As discussed in Section 5.4.1, temperature increases in the sludge are largely dependent on the water content; larger samples will require longer contact time to achieve the lethal temperature necessary for complete reduction. Despite 70°C being considered the minimum temperature lethal to pathogenic bacteria, live cells were still detected when this temperature was attained in the 20 g sample irradiated at 1,550 W for 10 seconds (see Figure 5-7a). This implies that upon attaining the minimum lethal temperature, the microorganisms need to be exposed for an additional minimum duration to ensure complete die off. Therefore, when the aim of treatment is only to destroy bacteria in the sludge, the total contact time for the entire process will be the sum total of the time taken to attain the lethal temperature level and the additional minimum exposure time required for complete die off. How fast this is achieved will depend on the MW power level applied and the amount of material treated. Furthermore, it is evident from the results here that complete bacteria (*E. coli*) reduction can be achieved over short contact times when the material is irradiated at higher MW power levels. High power input results in rapid escalation of the temperature which is a key contributor to bacterial die off. Thus when targeting complete reduction in a MW based reactor, it is desirable to use high MW power to realize shorter retention time and smaller reactor footprints. This is desirable especially in the emergency situations where land space is often a major constraint. Moreover, comparing the MW energy requirements for weight/volume reduction and the *E. coli* reduction, it can be inferred that more power is required to realize substantial drying than the bacterial reduction. In the 100 g sample, for instance, while over 180 Wh was required to attain a weight/volume reduction of approximately 70% (Figure 5-5b), complete *E. coli* reduction was achieved at 23 Wh (Figure 5-6b).

5.4.4 Organic stabilization

From the results shown in Figures 5-8a and 5-8b, it is evident that MW treatment was not successful in VS/TS reduction in both the 20 g and 100 g sample. After the MW treatment, the final VS/TS ratios attained in each sample size were higher than the 60% recommended by the European Environment Agency (Bresters et al., 1997) as the reference for the organically stable sludge. Thus, under the test conditions carried out in this study, organic stabilization of the blackwater FS was not achieved. This is expected as the highest temperature attained during the irradiation was approximately 127°C (i.e. 20 g sample, 1,550 W, four minutes contact time) which is much lower than the 550°C normally applied for VS ignition in the gravimetric method (SM-2540E) (APHA, 2012) during the VS measurements. Nevertheless, if the MW technology is aimed at specific situations such as the FS treatment in areas of intensive sanitation facility usage, e.g. emergency situations, the organic sludge instability may not be a major concern provided that pathogens are fully inactivated and the public health risk is reduced. Furthermore, as demonstrated in this study, the sludge weight/volume can be largely reduced by the MW irradiation making it possible to cost-efficiently treat it further (for organic stabilization) by less costly options such as composting, sludge drying beds, anaerobic digestion, etc., that also promote resource recovery.

5.5 Conclusions and recommendations

The MW treatment was not only able to achieve over 70% weight/volume reduction but also a complete reduction of the pathogenic bacteria indicator (*E. coli*) in the sludge. However, under the experimental conditions evaluated in this study, the MW treatment expectedly, did not yield substantial organic stabilization of the sludge. Nevertheless, with further developments the MW technology can be considered a promising option for the rapid treatment of fresh FS. Particularly, testing for more resistant microorganisms such as helminth (e.g. *Ascaris lumbricoides*) eggs, enteroviruses, etc., is desired to assess the MW capability to achieve complete FS sanitization. Further research is also required to develop a pilot-scale MW based reactor unit and test with FS from intensively used sanitation facilities such as toilets in emergencies or similar situations.

Acknowledgments

This research was funded by the Bill and Melinda Gates Foundation under the framework of SaniUP project 'Stimulating Local Innovation on Sanitation for the Urban Poor in Sub-Saharan Africa and South-East Asia' (OPP1029019).

References

APHA, 2012. Standard Methods for the Examination of Water and Wastewater, 22nd ed. American Public Health Association, Washington DC, USA.

Banik, S., Bandyopadhyay, S., Ganguly, S., 2003. Bioeffects of microwave—a brief review. Bioresour. Technol. 87, 155-159.

Bennamoun, L., Arlabosse, P., Léonard, A., 2013. Review on fundamental aspect of application of drying process to wastewater sludge. Renew. Sust. Energ. Rev. 28, 29-43.

Border, B.G., Rice-Spearman, L., 1999. Microwaves in the laboratory: Effective decontamination. Clin. Lab. Sci. 12, 156-160.

Bresters, A.R., Coulomb, I., Matter, B., Saabye, A., Spinosa, L., Utvik, Å.Ø., 1997. Management Approaches and Experiences: Sludge Treatment and Disposal., Environmental Issues Series 7. European Environment Agency, Copenhagen, Denmark, p. 54.

Byamukama, D., Kansiime, F., Mach, R.L., Farnleitner, A.H., 2000. Determination of Escherichia coli Contamination with Chromocult Coliform Agar Showed a High Level of Discrimination Efficiency for Differing Fecal Pollution Levels in Tropical Waters of Kampala, Uganda. Appl. Environ. Microbiol. 66, 864-868.

Eskicioglu, C., Terzian, N., Kennedy, K.J., Droste, R.L., Hamoda, M., 2007. Athermal microwave effects for enhancing digestibility of waste activated sludge. Water Res. 41, 2457-2466.

Fidjeland, J., Magri, M.E., Jönsson, H., Albihn, A., Vinnerås, B., 2013. The potential for self-sanitisation of faecal sludge by intrinsic ammonia. Water Res. 47, 6014-6023.

Flaga, A., 2005. Sludge drying, in: Plaza, E., Levlin, E. (Eds.), Proceedings of Polish-Swedish seminars. Integration and optimisation of urban sanitation systems., Cracow, Poland.

91

Retrieved   from
http://www2.lwr.kth.se/Forskningsprojekt/Polishproject/index.asp?entry=13#Report13
.

Gan, Q., 2000. A case study of microwave processing of metal hydroxide sediment sludge from printed circuit board manufacturing wash water. Waste Manage. (Oxford) 20, 695-701.

Haque, K.E., 1999. Microwave energy for mineral treatment processes—a brief review. Int. J. Miner. Process. 57, 1-24.

Hong, S.M., Park, J.K., Lee, Y.O., 2004. Mechanisms of microwave irradiation involved in the destruction of fecal coliforms from biosolids. Water Res. 38, 1615-1625.

Hong, S.M., Park, J.K., Teeradej, N., Lee, Y.O., Cho, Y.K., Park, C.H., 2006. Pretreatment of Sludge with Microwaves for Pathogen Destruction and Improved Anaerobic Digestion Performance. Water Environ. Res. 78, 76-83.

Ingallinella, A.M., Sanguinetti, G., Koottatep, T., Montanger, A., Strauss, M., 2002. The challenge of faecal sludge management in urban areas - strategies, regulations and treatment options. Water Sci. Technol. 46, 285-294.

Katukiza, A.Y., Ronteltap, M., Niwagaba, C.B., Foppen, J.W.A., Kansiime, F., Lens, P.N.L., 2012. Sustainable sanitation technology options for urban slums. Biotechnol. Adv. 30, 964–978.

Lamb, A.S., Siores, E., 2010. A Review of the Role of Microwaves in the Destruction of Pathogenic Bacteria, in: Anand, S.C., Kennedy, J.F., Miraftab, M., Rajendran, S. (Eds.), Medical and Healthcare Textiles. Woodhead Publishing, pp. 23-29.

Lin, Q.H., Cheng, H., Chen, G.Y., 2012. Preparation and characterization of carbonaceous adsorbents from sewage sludge using a pilot-scale microwave heating equipment. J. Anal. Appl. Pyrolysis 93, 113-119.

Lopez-Vazquez, C.M., Dangol, B., Hooijmans, C.M., Brdjanovic, D., 2014. Co-treatment of Faecal Sludge in Municipal Wastewater Treatment Plants, in: Strande, L., Ronteltap, M., Brdjanovic, D. (Eds.), Faecal Sludge Management - Systems Approach Implementation and Operation. . IWA Publishing, London, UK, pp. 177-198.

Menéndez, J.A., Domínguez, A., Inguanzo, M., Pis, J.J., 2005. Microwave-induced drying, pyrolysis and gasification (MWDPG) of sewage sludge: Vitrification of the solid residue. J. Anal. Appl. Pyrolysis 74, 406-412.

Menéndez, J.A., Inguanzo, M., Pis, J.J., 2002. Microwave-induced pyrolysis of sewage sludge. Water Res. 36, 3261-3264.

Oxfam, 2011. Urban WASH Lessons Learned from Post-Earthquake Response in Haiti, Oxford, UK., pp. 11. Retrieved from http://policy-practice.oxfam.org.uk/publications/urban-wash-lessons-learned-from-post-earthquake-response-in-haiti-136538.

Remya, N., Lin, J.-G., 2011. Current status of microwave application in wastewater treatment—A review. Chem. Eng. J. 166, 797-813.

Richard, C., 2001. Excreta-related infections and the role of sanitation in the control of transmission, in: Lorna, F., Jamie, B. (Eds.), Water Quality: Guidelines, Standards & Health: Assessment of Risk and Risk Management for Water-Related Infectious Disease. IWA Publishing, London, UK., pp. 89-113.

Ronteltap, M., Dodane, P.-H., Bassan, M., 2014. Overview of Treatment Technologies, in: Strande L, Ronteltap M, Brdjanovic D (Eds.), Faecal Sludge Management - Systems Approach Implementation and Operation. IWA Publishing, London, UK, pp. 97-120.

Sangadkit, W., Rattanabumrung, O., Supanivatin, P., Thipayarat, A., 2012. Practical coliforms and Escherichia coli detection and enumeration for industrial food samples using low-cost digital microscopy. Procedia Eng. 32, 126-133.

Shamis, Y., Taube, A., Shramkov, Y., Mitik-Dineva, N., Vu, B., Ivanova, E.P., 2008. Development of a microwave treatment technique for bacterial decontamination of raw meat. International Journal of Food Engineering 4.

Strande, L., 2014. The Global Situation, in: Strande, L., Ronteltap, M., Brdjanovic, D. (Eds.), Faecal Sludge Management - Systems Approach Implementation and Operation. . IWA Publishing London, UK, pp. 1-14.

Tang, B., Yu, L., Huang, S., Luo, J., Zhuo, Y., 2010. Energy efficiency of pre-treating excess sewage sludge with microwave irradiation. Bioresour. Technol. 101, 5092-5097.

Tappero, J.W., Tauxe, R.V., 2011. Lessons Learned during Public Health Response to Cholera Epidemic in Haiti and the Dominican Republic. Emerging Inefectious Diseases 17, 2087–2093.

Thostenson, E.T., Chou, T.W., 1999. Microwave processing: fundamentals and applications. Compos. A: Appl. Sci. Manuf. 30, 1055-1071.

Tyagi, V.K., Lo, S.-L., 2012. Enhancement in mesophilic aerobic digestion of waste activated sludge by chemically assisted thermal pretreatment method. Bioresour. Technol. 119, 105-113.

Tyagi, V.K., Lo, S.-L., 2013. Microwave irradiation: A sustainable way for sludge treatment and resource recovery. Renew. Sust. Energ. Rev. 18, 288-305.

Valero, A., Cejudo, M., García-Gimeno, R.M., 2014. Inactivation kinetics for Salmonella Enteritidis in potato omelet using microwave heating treatments. Food Control 43, 175-182.

Watson, J.T., Gayer, M., Connolly, M.A., 2007. Epidemics after Natural Disasters. Emerging Infectious Diseases 13, 1-5.

WHO, 2006. Wastewater and Excreta use in Aquaculture, Guidelines for the safe use of wastewater, excreta and greywater. WHO, Geneva, Switzerland, p. 219.

Wojciechowska, E., 2005. Application of microwaves for sewage sludge conditioning. Water Res. 39, 4749-4754.

Yu, Q., Lei, H., Li, Z., Li, H., Chen, K., Zhang, X., Liang, R., 2010. Physical and chemical properties of waste-activated sludge after microwave treatment. Water Res. 44, 2841-2849.

Chapter 6
Microwave treatment of faecal sludge from intensively used toilets in the slums of Nairobi, Kenya

This chapter is based on:

Mawioo, P.M., Garcia, H.A., Hooijmans, C.M., Brdjanovic, D., 2016. Microwave treatment of faecal sludge from intensively used toilets in the slums of Nairobi, Kenya. Journal of Environmental Management 184, Part 3, 575-584.

Abstract

Toilet facilities in highly dense areas such as the slum and emergency settlements fill up rapidly; thus, requiring frequent emptying. Consequently, big quantities of fresh faecal sludge (FS) containing large amounts of pathogens are generated. Fast and efficient FS treatment technologies are therefore required for safe treatment and disposal of the FS in such conditions. This study explores the applicability of a microwave (MW) technology for the treatment of fresh FS obtained from urine-diverting dry toilets placed in slum settlements in Nairobi, Kenya. Two sample fractions containing 100 g and 200 g of FS were exposed to MW irradiation at three input MW power levels of 465, 1,085 and 1,550 W at different exposure times ranging from 0.5 to 14 minutes. The variation in the FS temperature, pathogen reduction via the destruction of *E. coli* and *Ascaris lumbricoides* eggs, and volume/weight reduction were measured during the MW treatment. It was demonstrated that the MW technology can rapidly and efficiently achieve complete reduction of *E. coli* and *Ascaris lumbricoides* eggs, and over 70% volume/weight reduction in the fresh FS. Furthermore, the successful evaluation of the MW technology under real field conditions demonstrated that MW irradiation can be applied for rapid treatment of fresh FS in situations such as urban slum and emergency conditions.

6.1 Introduction

Sanitation facilities, especially the toilets provided in densely populated areas, such as urban slums and emergency settlements, fill up fast due to intensive use and they require frequent emptying. For instance, Sanergy, a social enterprise, empties fresh faecal sludge (FS) from over 700 toilet units that serve over 30,000 users daily in the informal settlements of Nairobi, Kenya (Sanergy, personal communication). Also, approximately 50 - 200 users per toilet per day are commonly observed in disaster situations (The Sphere Association, 2018); especially, at the onset of emergencies (UNHCR, personal communication). Consequently, large quantities of fresh FS are generated which require safe treatment and disposal. Various issues are identified that generally present a challenge to the FS management, especially in densely populated conditions. FS contains high amounts of pathogens such as bacteria, helminths, viruses, protozoa, and others (Richard, 2001; Jimenez et al., 2006; Fidjeland et al., 2013), which can pose a great risk to the public heath if it is inappropriately managed. In addition, the large amount of emptied fresh FS may need to be transferred to disposal sites far away from the points of generation, as in slum and emergency settlements where land space constraints and lack of adequate disposal possibilities are common. Massive expenditure may thus be incurred in emptying and transporting large amounts of FS, which would make the operation and maintenance of the sanitation system overly expensive. An example is the Haiti emergency camps six months after the 2010 earthquake, where a relief agency (Action Contre la Faim (ACF)) still incurred a monthly expenditure of approximately USD 500,000 to empty the toilets and dispose of FS (Bastable and Lamb, 2012). FS also contains high amounts of organic matter whose uncontrolled degradation in the environment can result in the generation of offensive odour, which may cause respiratory-related complications and attract disease vectors. These concerns form a major challenge to FS management in densely populated areas; hence requiring solutions that are more adapted to those conditions. The common sanitation solutions provided in urban slum and emergency settlements are mainly containment options. These comprise a range of onsite toilet facilities including chemical toilets, packet toilets (e.g. peepoo and wagbag), bucket latrines or elevated toilets, trench latrines, pit latrines, and others (Harvey, 2007; Katukiza et al., 2012). In recent years, there have also been remarkable efforts to expand the onsite toilet options in which a number of prototypes have been developed. However, parallel efforts to develop technology options to treat FS generated from those toilets still have to be demonstrated in practice.

Various treatment alternatives are available for FS such as composting, co-composting with organic solid waste, conventional drying (e.g. in sludge drying beds), anaerobic co-digestion with organic solid waste, and co-treatment in wastewater treatment plants (Ingallinella et al., 2002; Katukiza et al., 2012; Ronteltap et al., 2014). However, they are mostly suited for regular sanitation contexts and have limitations such as slow treatment processes, large land space requirements, among others, (Mawioo et al., 2016) which hinder their application to situations with unusually high rate of FS generation. Consequently, there is need to develop FS treatment technologies that are more appropriate for conditions such as those prevailing in slums and emergencies. Key among the desired characteristics for an appropriate FS treatment technology in these situations is that it should be fast, efficient, and compact for easy and rapid deployment.

Particularly, the technology should as much as possible address the various issues of concern mentioned above. In those areas with a high generation rate of fresh FS (e.g. slums and emergencies), the reduction of pathogenic organisms should definitely be prioritized over sludge volume and organic matter reduction so as to minimize the risk of excreta-related disease outbreaks. The amount of the pathogenic organisms should be reduced to the recommended safe levels (e.g. *E.coli* to $\leq 1.0 \times 10^3$ CFU/g TS and *Ascaris* eggs to < 1 *Ascaris* egg/g TS (WHO, 2006)). Next to pathogen reduction, it is often desirable to reduce the FS volume (to minimize handling costs) and organic matter content (to avoid odour and disease vector nuisance).

A microwave (MW) based technology can be a viable option for the treatment of fresh FS from intensively used toilet facilities as it has been shown being regarding the efficient pathogen inactivation and volume reduction (Mawioo et al., 2016). MW irradiation uses the MW energy (E_{MW}) with wavelengths between 1mm and 1m and frequencies between 300 MHz and 300 GHz in the electromagnetic spectrum (Haque, 1999; Tang et al., 2010; Remya and Lin, 2011). The MW technology has been used in various applications, most of which involve the use of heat generation. In such applications, the heat is generated by the molecular motion in the target material resulting from the migration of ionic species and/or rotation of the dipolar species when they interact with the microwaves (Haque, 1999; Thostenson and Chou, 1999; Mawioo et al., 2016). Various benefits are associated with heating by MW (Haque, 1999; Mawioo et al., 2016). The heating of a material by microwaves depends on its dielectric properties (i.e. the dielectric loss factor and the dielectric constant) and materials with high dielectric loss factor are favourable for the MW heating. Various types of sludge, such as sewage sludge and blackwater sludge (i.e. sludge extracted from a blackwater stream generated in low flush toilets, TS = 12%) contain dipolar molecules (e.g. water and organic complexes) with high loss dielectric properties and have demonstrated a good response to the MW treatment (Yu et al., 2010; Mawioo et al., 2016). For instance, nearly complete bacterial removal was reported when sewage sludge (Hong et al., 2004; Hong et al., 2006) and blackwater sludge (Mawioo et al., 2016) were heated by MW to temperatures above 65 °C. Furthermore, over 70% sludge volume reduction was achieved by treating blackwater sludge (Mawioo et al., 2016) and anaerobic sewage sludge (Menéndez et al., 2002) with MW. The MW effect on the pathogen destruction is linked to both the non-thermal (electromagnetic radiation) and thermal (temperature) effects of electromagnetic energy (Banik et al., 2003; Hong et al., 2004; Mawioo et al., 2016). By electromagnetic radiation, molecules of the irradiated material orient themselves in the direction of the electric field, which may break the hydrogen bonds leading to the denaturation and death of microbial cells (Banik et al., 2003; Tyagi and Lo, 2013). Conversely, the destruction by thermal effect is caused by the rapturing of microbial cells when water is rapidly heated to the boiling point by rotating dipole molecules under an oscillating electromagnetic field (Hong et al., 2004; Tang et al., 2010; Tyagi and Lo, 2013). On the other hand, volume reduction is strongly linked to the temperature increase, which causes evaporation of the water contained in the sludge (Mawioo et al., 2016). Stabilization of organic matter in sludge was not achieved by MW heating, arguably due to the relatively low maximum temperature (i.e. 127 °C) attained in the treatment process (Mawioo et al., 2016). However, organic stabilization was achieved by Menéndez et al. (2002) when they mixed sludge with a MW receptor material to attain high temperatures (over 900 °C).

As discussed above, waste management by MW technology has been demonstrated through treating the various kinds of sludge. The technology possesses a rapid heating and treatment capability, which can be explored further for possible applications in treating fresh FS in slum and emergency settlements. Despite the reported successes in the various kinds of waste treatment, the information on the evaluation of the MW technology in FS treatment for a potential field application is still limited. So far, only a recent study evaluating the MW treatment of blackwater sludge has been reported; in which *E. coli*, sludge volume, and organic matter reduction was assessed (Mawioo et al., 2016). It is therefore highly needed to evaluate the potential application of this technology for specific field applications using fresh FS obtained from toilet facilities under real field conditions. In addition, the previous studies demonstrated pathogen reduction on *E.coli* and faecal coliforms; therefore, including other pathogenic organisms such as helminth eggs is also important, as they are shown to be more resistant to treatment (Feachem, 1983; WHO, 2006; Koné et al., 2007). If successful, the information derived from such study can help to further validate the MW application for treatment of FS under those conditions and set the basis for scaling up the technology. Also, the study can expand knowledge about the response of different types of sludge to MW treatment.

In this study the potential of a MW based technology for slum or emergency sanitation applications was evaluated by treating fresh FS obtained from toilets in the slums of Mukuru and Mathare in Nairobi, Kenya. Besides being a representation of FS in an urban slum environment, the fresh FS sample obtained in these conditions is also relatively similar to that which is generated in emergency camps. Three aspects of the proposed MW treatment technology were assessed in this research including the reduction of pathogens, sludge volume, and organic matter. Both *E. coli* and helminth (*Ascaris Lumbricoides*) eggs were used as indicators for pathogen reduction, while the sludge weight was used to estimate the volume reduction. The organic stabilization of the FS was estimated using the volatile and total solids ratio (VS/TS) as indicator.

6.2 Materials and methods

6.2.1 Research design

This study was performed using fresh FS samples obtained from Fresh Life® toilets (Figure 6-1), which are installed and maintained by Sanergy in collaboration with entrepreneurs in the slums of Nairobi, Kenya.

Figure 6-1. Fresh Life® toilets installed in a slum in Nairobi, Kenya

Fresh Life® is the brand name of Sanergy toilets that uses the principle of urine-diverting dry toilet (i.e. faeces and urine streams are diverted and collected in separate containers). The toilets are emptied on a daily basis by removing and replacing the filled containers with clean empty ones. Over seven metric tons of FS are then transported to a central treatment facility (approximately 30 kilometers from Nairobi), where they are converted into fertilizer via composting. The ultimate goal of this study was to test the MW technology for treatment of fresh FS generated from intensively used toilet facilities in areas such as slums and emergencies. Therefore, the Fresh Life® toilets were chosen for the FS source because besides their location in a slum settlement, they are also comparable to the emergency toilet facilities in many aspects. For instance, the dry (i.e. non flush) toilet systems, similar to the Fresh Life® toilets, are the most common FS management technologies applied in emergency camps (Harvey, 2007). Furthermore, due to the high population densities in the slum settlements, the intensive toilet usage and corresponding rapid fill up is similar to that encountered in emergency camps. These similarities allowed the study to be also relevant to the emergency conditions. All of the necessary research infrastructures (e.g. specialized FS laboratory and expert support) was made available to the research team through funding from the Bill & Melinda Gates Foundation.

During the study, two amounts of FS (i.e. 100 and 200 g) were treated by exposing it to MW irradiation in a domestic MW oven for various durations and input MW power levels. The changes in the various parameters including temperature, *E. coli*, helminths (*Ascaris lumbricoides)* eggs, weight, and VS/TS ratio were measured in the treated samples. The experiments for both the 100 g and 200 g samples were conducted in triplicates and repeated to obtain three trials. The effectiveness of the MW treatment was then determined by the changes in the measured parameters between the raw and the treated FS samples (Mawioo et al., 2016).

6.2.2 Microwave apparatus

A domestic microwave oven, Samsung, MX245 (Samsung Electronics Benelux B.V., the Netherlands) was used in this study.

The unit operates at a frequency of 2,450 MHz with a power output ranging from 0 to 1,550 W with 10% incremental steps. The microwave oven was placed in a makeshift structure that was located at the Sanergy's central waste treatment facility in Nairobi, Kenya where the entire research activities were carried out.

6.2.3 FS samples

Containers with FS from three toilets were identified from which approximately equal, but large, portions of the fresh FS samples were obtained and transferred into a plastic bucket. The samples were then mixed thoroughly on site to attain a homogenous sample from which three smaller samples were drawn and placed into plastic sampling containers. The samples were then transported to the Sanergy's central waste treatment facility where the MW treatment experiments were conducted within 24 hours. The characteristics of the fresh FS are presented in Table 6-1.

Table 6-1. Characteristics of the raw FS, N = 6

Parameter	Average	STDEV
Water content, %	74	2
Total solids, %	26	2
VS/TS ratio	0.92	0.01
TCOD (mg O2/g TS)	1.98×10^5	2.8×10^4
E. coli (CFU/g TS)	4.0×10^8	8×10^7
Ascaris lumbricoides eggs (eggs/g TS)	Not detected	-

6.2.4 Experimental procedures

Sample preparation

The two sizes of test samples were prepared in triplicates: the 100 g FS samples were placed in one liter glass beakers (height of the FS was approximately one centimeter, surface area approximately 78.5 cm^2), while the 200 g samples were placed in two liter glass beakers (height of the FS was approximately one centimeter, surface area approximately 156 cm^2) (Figure 6-2).

Figure 6-2. The raw FS test samples in 1L and 2L beakers

As shown in Table 6-1 above, the *E. coli* naturally occurring in the FS (i.e. 4.0 x 10^8 CFU/g TS) was sufficient for evaluating the performance of the MW technology on *E. coli* inactivation; thus spiking *E.coli* was not necessary. However, an analysis of the FS samples did not reveal existence of *Ascaris lumbricoides* (helminths indicator) eggs. Therefore, the evaluated FS samples were spiked with *Ascaris lumbricoides* eggs before the MW treatment. The samples of 100 g and 200 g were spiked by adding and mixing approximately 1 x 10^4 and 2 x 10^4 *Ascaris lumbricoides* eggs, respectively.

Microwave treatment

The FS samples were treated in the MW apparatus (Section 6.2.2). The sample contained in the glass beaker was placed in the MW cavity and then irradiated at 465, 1,085, and 1,550 W for varied time durations (i.e. between 0.5 and 14 minutes). After the MW treatment, the sample was removed from the MW cavity (Figure 6-3) and its temperature was immediately measured before covering with sanitized aluminum foil.

Figure 6-3. The MW treated FS test samples in 1L beakers

The samples that underwent microwave treatment were cooled down to room temperature and analyzed for their characteristics as described in the following sections.

6.2.5 Analytical procedures

FS samples with and without MW treatment were measured for various physical-chemical parameters including temperature, weight, TS, and VS and the microbial parameters including the *E. coli* and *Ascaris lumbricoides* eggs.

COD measurement

A known amount of FS sample (prior to MW treatment) was diluted in demineralized water, after which the COD concentration was measured according to the closed reflux method (SM 5220 C) (APHA, 2012) and expressed in mg COD per g TS (mg COD/g TS) (Table 6-1).

Temperature

The initial sample temperature was measured just before MW treatment using an infrared thermometer (Fluke 62 MAX, Fluke Corporation, U.S.A). Following each treatment, the sample was taken out from the MW cavity and the temperature was immediately measured. Due to the solid nature of the sample, the temperature was only measured on the surface.

Weight measurement

The initial weight was measured using a bench-top weighing balance (Sartorius H160, Sartorius AG, Germany) as the samples were transferred into the heating beakers. The final weight after MW treatment was measured once the samples were cooled to room temperature. The volume reduction was then determined from the initial and the final sample weight difference. Based on the maximum temperature attained during MW treatment (i.e. $\leq 134\,^{\circ}C$), the weight reduction could mainly be attributed to the water evaporating from the heated sludge. Thus, considering the density of water, the weight reduction was deemed to be equivalent to the sludge volume reduction.

TS and VS measurement

TS content was determined by drying the samples in an oven at $105\,^{\circ}C$ for at 24 hours (for TS), after which they were cooled and weighed. The VS was determined in the same samples by burning in a muffle furnace at $550\,^{\circ}C$ for two hours. The TS and VS results were then used to evaluate the organic stability of sludge.

E. coli measurement

The detection of *E. coli* was done using the surface plate technique with chromocult coliform agar (Chromocult; Merck, Darmstadt, Germany) (Byamukama et al., 2000). A step by step procedure similar to what is previously reported in Mawioo et al. (2016) was applied. Dark blue to violet colonies were classified as *E. coli* (Byamukama et al., 2000; Sangadkit et al., 2012).The average number of colonies were used to calculate the viable-cell concentrations in the samples, expressed in CFU/g TS of the test sample.

Recovery of *Ascaris* eggs and incubation

The *Ascaris eggs* recovery was done according to the protocol developed by Moodley et al. (2008) and modified in Pebsworth et al. (2012). Portions (20 g) of the respective MW treated samples were placed on a clean plastic beaker to which 80 mL of ammonium bicarbonate (119 g of ammonium bicarbonate in one liter of de-ionized water) was added and then mixed on a magnetic stirrer for 20 minutes. The mixture was poured through a 100-μm sieve fitted on top of a 25 μm sieve (200mm dia.) and thoroughly washed with tap water. The material retained on the 100 μm sieve was discarded while that retained on the 25-μm sieve was thoroughly washed and then rinsed into a clean plastic beaker. The solution was then transferred into 15 mL Falcon tubes and centrifuged at 3000 rpm for five minutes using a bench-top centrifuge (EBA 20, Andreas Hettich GmbH &CO. KG, Germany). The supernatant was discarded while the solid pellet containing the eggs) was re-suspended by adding $ZnSO_4$ (specific gravity 1.3) when vortexing until the tubes were filled to the 14mL mark. The solution was centrifuged at 2,000 rpm for five minutes and the resultant supernatant floatation fluid poured over a smaller 25 μm sieve (100mm dia.). The material retained on the sieve was washed well with tap water and then rinsed into a clean plastic beaker. The solution was transferred back into the 15 mL plastic test tubes (Falcon) and centrifuged one last time at 3,000 rpm for five minutes. The supernatant was discarded and the egg pellets (the concentrate of the centrifugation) were transferred into a 50 mL Falcon tube containing 10 mL of de-ionized water. The Falcon tube was covered with a plastic film (Parafilm) that was pricked (to allow air into the sample) and acted as a humid chamber for incubation at 28 °C for 28 days.

Viability test and eggs count

After the 28 days incubation, the eggs were transferred to the 15 mL Falcon tubes and centrifuged for five minutes at 3,000 rpm. The supernatant was removed and the remaining pellet containing eggs was well mixed using a pipette. The suspension of the eggs (1mL) was placed on a microscopic slide and covered with a cover slip. The slide was observed under the microscope (AmScope, California, USA) at a magnification of 10 and 40. The eggs were counted as living if they contained a fully developed larva and dead without a larva but with an internal structure.

6.2.6 Data analysis

The experimental data was processed using Microsoft Excel software. Firstly, the data for each trial were separately processed by computing the average values for each set of the triplicate treatments. Then the average values obtained in each of the three trials were further combined by computing their mean values. Furthermore, the respective standard deviations and standard errors for the combined trials mean values were calculated. The mean values for the three trials were then presented in either tables or figures with their respective standard error values or error bars shown. Linear regression was performed; particularly to estimate the specific energy demand rates (i.e. watt-hour, Wh per gram) on each of the three drying phases.

6.3 Results

6.3.1 Temperature evolution

The temperature profiles for the 100 g and 200 g sample during the MW treatment are shown in Figure 6-4a and 6-3b, respectively. Three distinctive temperature evolution phases were observed during the respective input MW power levels and contact times evaluated. Furthermore, different temperature propagation rates were observed among the three phases. For instance, there was a rapid rise in the FS temperature during the initial phase (Figure 6-4a and 6-1b) while a fairly constant and minimal rise was observed in the second phase.

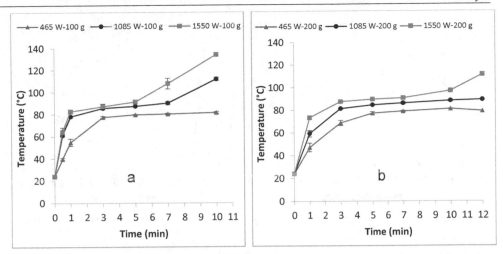

Figure 6-4 - Effect of exposure to microwaves on temperature evolution in a) 100 g sample and b) 200 g sample

A rapid rise in temperature was again observed during the third phase. It is also evident from Figure 6-4a and 6-3b that the temperature increment was more rapid in the 100 g than the 200 g sample in the initial phase. For instance, while after a one minute contact time the 100 g sample attained a temperature of 55, 78, and 83 °C at the respective input MW power levels of 465, 1,085, and 1,550 W; the corresponding temperatures attained in the 200 g sample were lower, namely 47, 59, and 73 °C, respectively.

6.3.2 Pathogen reduction

E. coli reduction

Figure 6-5a1 and 6-4a2, and 6-4b1 and 6-4b2 show the results of the *E. coli* reduction in the 100 and 200 g samples obtained at various input MW power levels and contact times, respectively. Furthermore, the influence of temperature on the *E.coli* reduction over contact time was observed. The results show increased *E. coli* reduction when the input MW power level and/or the contact time was increased. Also, as expected, there was an observed reduction of *E. coli* with the rise in the FS temperature. For instance, a reduction of *E. coli* was achieved when the 100 g sample was heated at 465 W for one minute (i.e. E_{MW} = 8 Wh, T = 55 °C) resulting to approximately 0.74 log removal value (LRV), i.e. 80% removal. However, an even higher reduction below the detection limit (i.e. <1.0x10^3 CFU/g TS or approximately 5.60 LRV) was achieved when the input MW power level was raised to 1,085 W (E_{MW} = 18 Wh, temp = 78 °C) and 1,550 W (26 Wh, temp = 83 °C) at the one-minute contact time. A reduction below the detection limit was achieved with the 465 W when the contact time was increased to three minutes (i.e. E_{MW} = 23 Wh, T = 77 °C).

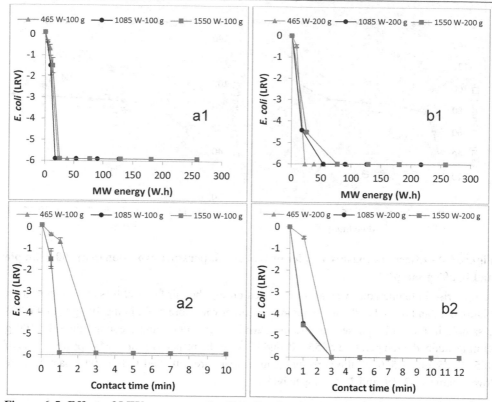

Figure 6-5. Effect of MW energy on the *E. coli* reduction in a1) 100 g FS sample and b1) 200 g FS sample, and *E. coli* reduction as a function of contact time in a2) 100 g FS sample and b2) 200 g FS sample. The zero *E.coli* log removal corresponds to an initial concentration of 4.0×10^8 CFU/g TS).

A similar trend to that observed in the 100 g sample was also demonstrated in the 200 g sample (Figure 6-5 b1 and 6-2b2) in which the increment in the input MW power led to increased *E. coli* reduction. However, lower inactivation efficiencies were observed in the 200 g sample compared to the 100 g sample. For instance, while reduction below the detection limit was achieved with the 1,085 W and 1,550 W at the one-minute contact time in the 100 g sample, only approximately 4.4 LRV was achieved in the 200 g sample. Nevertheless, *E. coli* reduction below the detection limit was achieved in each evaluated input MW power at the three minutes contact time. The corresponding energy and temperature levels attained at the three minutes contact time were 23 Wh and 69 °C, 54 Wh and 81 °C, and 78 Wh and 87 °C for the 465 W, 1,085 W and 1,550 W, respectively.

Ascaris egg reduction

Figure 6-6a1 and 6-6a2, and 6-6b1 and 6-6b2 presents the profiles for *Ascaris lumbricoides* egg reduction as a function of the MW energy and the exposure time for the 100 g and 200 g FS samples, respectively.

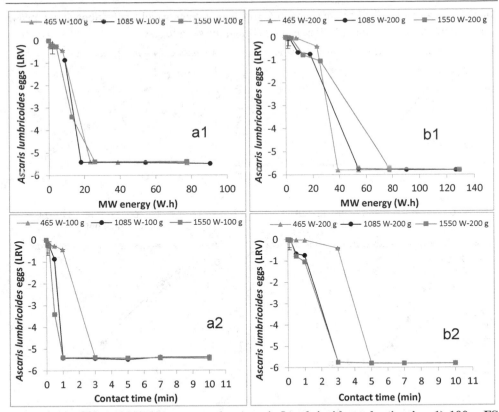

Figure 6-6. Effect of MW energy on the *Ascaris Lumbricoides* reduction in a1) 100 g FS sample and b1) 200 g FS sample, and *Ascaris Lumbricoides* reduction as a function of contact time in a2) 100 g FS sample and b2) 200 g FS sample. The zero *Ascaris Lumbricoides* eggs log removal corresponds to an initial concentration of 2.69 x 10^2 *Ascaris* eggs /g TS

The influence of temperature on the *Ascaris* egg reduction over contact time was also observed. As shown in the figure, the MW treatment was successful in achieving over 3 LRV reduction of the *Ascaris* eggs within one minute when the 100 g sample was exposed to MW irradiation at 1,085 W (E_{MW} = 18 Wh, T = 78 °C) and 1,550 W (E_{MW} = 26 Wh, T = 83 °C). However, to achieve a similar reduction at 465 W, a longer contact time, at least three minutes (E_{MW} = 23 Wh, T = 77 °C) was required. On the other hand, for the 200 g sample, at least a three minute contact time was required for the 1,085 W (i.e. E_{MW} = 54 Wh, T = 81 °C) and 1,550 W (i.e. E_{MW} = 78 Wh, T = 87 °C) to achieve over 3 LRV reduction. However, a longer contact time (at least five minutes, i.e. E_{MW} = 39 Wh, T = 77 °C) was required to achieve similar results when 465 W was used.

6.3.3 Volume reduction and energy requirements

Figure 6-7a and 6-6b presents the profiles for weight/volume reduction in the two sample fractions evaluated. The resulting trends are strongly linked to the temperature evolution and the three phases described above (Section 6.3.1). The volume/weight reductions at any phase varied between both the input MW power levels and the sample sizes.

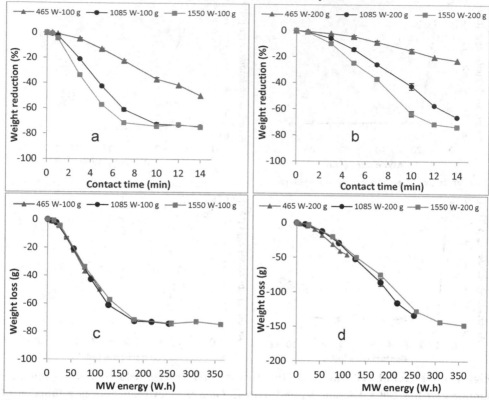

Figure 6-7. Effect of exposure to MW irradiation on sludge weight

The highest weight reductions at any phase in the two samples were achieved with the highest input MW power evaluated (i.e. 1,550 W), while the least was at the lowest input power (i.e. 465 W). In the 100 g sample, for instance, the initial phase that lasted up to one minute achieved a weight reduction of 1.0, 2.3, and 4.5% for the 465, 1,085 and 1,550 W, respectively. On the other hand, respective weight reductions of 0.8, 1.1, and 1.4% for the 465, 1,085 and 1,550 W were attained in the 200 g sample within the one-minute contact time (i.e. initial phase). In all cases of sample fractions, substantial moisture evaporation appears to start in the subsequent second phase exhibited by a high but relatively constant weight reduction rate. The lowest volume/weight reduction for both sample fractions was observed in the final phase, which was attained in the 100 g at 1,085 W and 1,550 W with contact time beyond 10 minutes and seven minutes, respectively. However, this phase was only achieved with the 1,550 W in the 200 g sample with contact time beyond 10 minutes.

In order to determine the energy demand during the MW treatment, the energy consumption profiles were observed (Figure 6-7c and 6-6d). Like the temperature evolution and weight reduction, the energy demand profiles correspond with the three phases observed. The energy demand results (estimated by computing linear regressions on the three phases) demonstrated relatively high but varied energy demands between the two sample fractions, especially during the initial phase. For instance, the 200 g sample exhibited a higher energy demand (approximately 7 Wh per gram of weight loss or 7 kWh per kg) than the 100 g sample (approximately 6 Wh per gram or 6 kWh per kg).

6.3.4 Total and volatile solids

The VS and TS for the 100 g and 200 g sample were measured and their respective VS/TS ratio computed. From the results there was no observed change in the VS/TS ratio, which was used as an indicator to estimate the organic stability in the treated FS. For each sample fraction and the input MW power levels tested, the final VS/TS ratio was between 86 and 92%.

6.4   Discussion

6.4.1 Temperature evolution

The three distinctive temperature propagation phases observed in Figure 6-4a and 6-3b are similar to those reported in the previous studies in which other types of sludge were treated using MW (Mawioo et al., 2016) and other drying methods such as convection and conduction (Flaga, 2005; Bennamoun et al., 2013). The observed initial, second, and final phases were classified as the preliminary drying phase, the essential (major) drying phase, and the final drying phase, respectively. The rapid temperature rise during the initial (preliminary) drying phase corresponds to that observed by Mawioo et al. (2016) and Yu et al. (2010) and is attributed to the high amounts of heat generated as a result of the interaction between the microwaves and the initially high concentration of the dipolar molecules (e.g. water, proteins, etc.) that are present in the wet sludge. The variability in temperature increments between the 100 g and the 200 g samples during the preliminary drying phase can be attributed to their respective initial water content. Water has a high thermal capacity and constitutes over 70% of the FS used in this study, thus the majority of the initial heat is absorbed in the water fraction of the FS. Hence, a lower temperature rise is expected in the 200 g sample, which has a higher water content (i.e. higher thermal absorption capacity) than the 100 g sample. Similar observations were reported when blackwater sludge and excess sewage sludge were heated by MW (Tang et al., 2010; Mawioo et al., 2016). The fairly constant and minimal temperature rise observed during the essential (i.e. second or major) drying phase was attributed to the possibility that the sludge may have reached the boiling point. During this phase, the unbound water is constantly evaporated from the surface of the sludge particles while being replaced by that from inside the particles (Flaga, 2005; Mawioo et al., 2016). The final (third) drying phase, as manifested in a rapid rise in the sludge temperature is a result of a more rapid evaporation of the water on the surface than it is replaced from the inside of sludge particles (Mawioo et al., 2016). Furthermore, it is evident from the results that the rate of temperature propagation is dependent both on the amount of sludge used and the input E_{MW} applied. Besides, the nature of material irradiated has influence on its temperature propagation behavior (Mawioo et al., 2016). This was demonstrated when a MW receptor material was mixed with sewage sludge to rapidly attain over 900 °C (Menéndez et al., 2002; Menéndez et al., 2005).

Generally, the results demonstrate that the MW irradiation is effective and fast in heating the FS. Previous studies demonstrated the temperature evolution rate was more rapid when material

109

is heated with MW than the conventional heating methods (Menéndez et al., 2002; Hong et al., 2004). Therefore, if applied for the FS treatment with the temperature as the main driver for the process, the MW technology can achieve a higher throughput and smaller reactor footprint. This can greatly address the challenge of land space constraints that often affect sanitation provision in scenarios such as the urban slum and emergency settlements.

6.4.2 Pathogen reduction

E. coli reduction

Generally, the results in Figure 6-5 show that MW treatment can provide a rapid and effective solution for the reduction of the *E. coli* in FS. This observation has also been reported in the previous studies (Border and Rice-Spearman, 1999; Lamb and Siores, 2010; Mawioo et al., 2016), but using different media than the fresh FS, which is used in this study. The increased reduction of the *E.coli* when either the input power and/or contact time were increased is due to the resulting rise in the E_{MW}, which ultimately is a key factor in the MW treatment process.

Various studies have attributed both the non-thermal (electromagnetic radiation) and the thermal (temperature) effects of the MW treatment to the destruction of microorganisms including the bacteria (Banik et al., 2003; Hong et al., 2004; Shamis et al., 2008; Tyagi and Lo, 2012; Valero et al., 2014). However, the thermal effect has been reported as the main mechanism for the destruction of bacteria with the minimum temperature for complete destruction identified at 70 °C (Hong et al., 2004; Tyagi and Lo, 2012; Valero et al., 2014; Mawioo et al., 2016). As shown in the results, generally the MW treatment achieves practically complete *E. coli* reduction within a short time duration. However, the contact time for complete bacterial reduction at given input MW power level will depend on the amount of FS treated. For instance, while *E. coli* was still detected in the 200 g FS sample when treated at 1,550 W and 1,085 W for one minute, it was not detected in the 100 g FS sample. Furthermore, the contact time for complete bacteria (*E. coli*) reduction for a given FS quantity can be substantially shortened when the irradiation is performed at high input MW power levels. It is also demonstrated in this study that despite the FS attaining the minimum lethal temperature for bacterial destruction, (i.e. 70 °C) *E. coli* can still be detected, for instance, when the 200 g sample was exposed 1,550 W at the one minute contact time (i.e. E_{MW} = 26 Wh, T = 73 °C). This observation agrees with that from a previous study by Mawioo et al. (2016), and shows that when the temperature is rapidly escalated to the lethal level, a minimum holding time is necessary to ensure compete destruction.

Ascaris egg reduction

Results from Figure 6-6 a1 and 6-5a2, and 6-5b1 and 6-5b2 demonstrate that the viability of *Ascaris lumbricoides* eggs in the FS can rapidly and effectively be reduced by the MW treatment. In this study, an energy dose of 18 Wh and 39 Wh was sufficient to produce sludge complying with the WHO guidelines of one *Ascaris* egg/g TS (i.e. approximately 2.4 LRV in this case) (WHO, 2006) in the 100 g and 200 g samples, respectively. These results agree with a previous study by Mun et al. (2009), who achieved a complete reduction of the *Ascaris lumbricoides* after a 60 second contact time at 700 W input MW power. However, they used an

Ascaris spiked soil media with smaller sample sizes (i.e. 25 g), which is different from the fresh FS used in this study. Furthermore, an approximate 2.2 log inactivation of *Taenia taeniaeformis* eggs was achieved when a 10 g sample of cat faeces was exposed to the MW treatment for only 18 seconds (Conder and Williams, 1983). However, it is not clear which input MW power level was evaluated in the study. Also, it was indicated that the *Taenia taeniaeformis* eggs might be three times less resistant to the MW treatment than the *Ascaris lumbricoides* eggs (Mun et al., 2009).

Like other microorganisms, the destruction of the *Ascaris* eggs may be attributed to both non-thermal (electromagnetic) and thermal effects during the MW treatment. Similar to the *E. coli* reduction, the thermal effect appears to play the major role in the destruction as demonstrated by the significant *Ascaris* egg reduction with the temperature rise in the FS samples. According to the current study, the lethal temperature to achieve a substantial destruction of the *Ascaris* egg appears to be approximately above 70 °C. This is in agreement with a previous study that used shear viscous heating in which over 90% helminth egg reduction was reported at 70 °C (Belcher et al., 2015). The results from the two sample fractions in this study suggest that the contact time required to attain the lethal temperature for the *Ascaris* egg inactivation varies with the FS quantity and input MW power applied. For the scenarios with land space constraints, as in in emergency settings, it is reasonable to choose the highest possible input MW power level so as to achieve a compact system with a small footprint. A comparison of the reduction results suggest more or less equal MW energy requirements for the destruction below the detection limit of *Ascaris lumbricoides* eggs (Figure 6-6 a1 and b1) and *E. coli* Figure 6-5 a1 and b1).

The short contact times for the reduction achieved with the MW treatment demonstrated that the technology is much more rapid in *Ascaris egg* inactivation than the commonly applied conventional treatment methods, such as the sludge drying beds, composting and co-composting processes, airtight storage, and others, which take over a month to achieve complete reduction (Feachem, 1983; Jimenez et al., 2006; Koné et al., 2007). However, the MW technology is considered more energy intensive than those conventional alternatives.

6.4.3 Volume reduction and energy requirements

The three phases observed in the volume/weight reduction (Figure 6-7a and 6-6b) and energy profiles (Figure 6-7c and 6-6d) are described above (Section 6.3.1). Results from Figure 6-7a and 6-4b demonstrate a low volume/weight reduction during the preliminary drying phase. This can be explained by the fact that the initial MW energy supplied is mainly used to heat the FS to the boiling point before the evaporation can start. On the other hand, a higher volume/weight reduction was achieved during the essential (major) drying phase since almost all the energy supplied at this stage is used for the evaporation of the unbound water that has low energy requirements (Mawioo et al., 2016). Furthermore, as reported in Mawioo et al. (2016), the duration of the entire essential drying phase varied with both the power input levels and the sample sizes. For instance, up to seven minutes in the 100 g sample and up to 10 minutes in the 200 g sample was required to conclude the essential drying when both samples were irradiated at 1,550 W. Shorter durations were also realized when the input MW power was raised within the same sample size. Therefore, the quantity of the irradiated sludge and input MW power are key determinants of the duration for the essential (major) drying phase. The lowest moisture loss observed during the final drying phase is due to the fact that the free water is completely exhausted at the essential drying stage. Therefore, any moisture loss at this point is only possible by evaporating the bound water requiring more energy, which explains the low weight reduction (Mawioo et al., 2016). These trends correspond to those observed when dewatered sediment sludge and blackwater sludge were subjected to MW drying (Gan, 2000; Mawioo et al., 2016).

The current study and other previous studies (Koné et al., 2007; Kengne et al., 2008; Fidjeland et al., 2013; Mawioo et al., 2016) have shown that the water fraction constitutes over 70% of the FS weight. Consequently, evaporation of water contributes to a greater extend the FS volume/weight reduction during the drying. For example, in this study, the over 70% overall FS volume/weight reduction can strongly be linked with evaporation of water during the entire MW treatment process. As shown in the results, water is mainly evaporated during the essential drying, which can be considered the most crucial to FS volume/weight reduction. This observation corresponds with previous studies on MW treatment of various sludge types (Menéndez et al., 2002; Mawioo et al., 2016). Therefore, the optimal operation window for an FS drying MW unit can be established within the essential (major) drying phase. In the practical field applications, the level of drying can be determined based on the cost of transporting the treated sludge to the disposal facilities.

Furthermore, the results from Figure 6-7c and 6-6d show a variation in energy demand between the two sample fractions that can be linked to their differences in total water content, which needs to be heated to the boiling point before any substantial moisture loss can be realized. The 200 g sample has a higher total water content and thus higher energy demand. The essential drying phase exhibited relatively similar energy consumption rates between the two sample fractions. As reported in Mawioo et al. (2016), the energy supplied in the essential phase is mostly used in the evaporation of water. The lowest energy demand is also exhibited in this phase, which is approximately 3 Wh per gram (3 kWh per kg) of weight loss for both sample fractions. This observation agrees with Mawioo et al. (2016) and can be explained by the evaporation of the free (unbound) water, which commences once the sludge is previously heated in the preliminary phase. Based on sewage sludge, Bennamoun et al. (2013) reported energy consumptions during essential drying between 0.7-1.4 and 0.8-1.0 kWh per kg of water evaporated in the convective and conductive industrial driers, respectively. The comparatively higher specific energy consumptions attained in this study can possibly be explained by the differences in the drying units' design and the sludge types treated, and the scale of treatment equipment (Mawioo et al., 2016). A microwave reactor with adequate ventilation and insulation can probably reduce the energy demand reported in the current study. The final drying phase, which was achieved with 1,085 W and 1,550 W in the 100 g sample and only 1,550 W in the 200 g sample, marked the highest sludge temperature and energy demand. In the case of the 100 g sample, for instance, energy demand of approximately 30 Wh was required per gram of weight loss (i.e. 30 kWh per kg). The removal of molecularly bound water at this stage requires more energy and thus the high energy demand.

Comparing the results for *E. coli* and *Ascaris* egg reduction with the volume/weight reduction, it can be observed that comparatively much higher MW energy is required to achieve substantial volume/weight reduction than that required for the complete pathogen reduction. For instance, while over 180 Wh was required to attain 70% FS volume/weight reduction, a complete *E. coli* and *Ascaris* egg reduction was achieved at 23 Wh in the 200 g sample. This suggests that when

FS drying is considered in the MW reactor design, then the pathogen reduction will also take place.

6.4.4 Total and volatile solids

Based on the results for VS and TS, there was no observed reduction of the VS/TS ratio in the two MW treated sample fractions. The final VS/TS ratio ranging between 86 and 92% was higher than the 60% recommended by the European Environment Agency (Bresters et al., 1997) as the reference for the stable sludge. A previous study by Mawioo et al. (2016) reported similar results in which they observed that the temperature attained with the blackwater sludge (i.e. 127 °C) was lower than the 550 °C normally recommended for VS ignition (APHA, 2012). Similarly, a maximum temperature of approximately 134 °C was attained in the current study (Figure 1a) which is considered very low for the VS ignition.

6.4.5 Microwave application for the treatment of faecal sludge from intensively used toilets

The results obtained in this study potentially demonstrate the technical feasibility of MW technology application for FS management in intensively used toilet facilities. The technology was found to be highly effective in the removal of both the *E. coli* and *Ascaris* eggs, which are considered most resistant in FS and are used as indicator organisms (Feachem, 1983; WHO, 2006; Koné et al., 2007). Relatively short contact time was required for pathogen reduction, implying that a short design retention time and small reactor footprint can be achieved. When FS drying is desired, however, expectedly longer retention times will be required depending on the initial water content in the FS and the level of drying desired. Additionally, the MW technology has relatively higher energy requirements than some conventional options (e.g. composting, anaerobic digestion, sun drying, etc.). Therefore, considering large-scale applications, the technology can be more appropriate for specific scenarios such as the urban emergencies where land space is limited and/or digging of pits is restricted leading to use of mobile toilets that require frequent emptying of fresh FS.

This study validates the MW application for the treatment of FS under the tested conditions and sets the basis for scaling up the technology and evaluating it on the basis of its performance, energy consumption, and operation. Economic analysis and comparison with other relevant technologies can also be performed. When designing a large-scale MW reactor unit for the practical field applications, a high capacity for input power (i.e. large microwave generators) should be considered to ensure rapid temperature escalations, which is important for the pathogen and sludge drying. This will ensure a smaller reactor footprint, which can easily be integrated into a compact (and if necessary) containerized system that is easy to transport and can quickly be erected and started up onsite.

6.5   Conclusions and recommendations

In this study, the evaluation of MW technology was conducted using fresh FS obtained from urine diversion toilets managed by a social enterprise, Sanergy, in Nairobi, Kenya. The fresh FS was obtained from toilets spread out in two informal settlements; Mukuru and Mathare slums of Nairobi. The study demonstrated for the first time that MW irradiation could be applied for rapid treatment of fresh FS obtained under slum conditions. Rapid temperature escalations can be achieved in the fresh FS, which is essential for sludge sanitization and volume reduction. In addition to the *E. coli*, *Ascaris lumbricoides* eggs, which are considered among the most resistant organisms in FS, were rapidly inactivated beyond the detection limit. A sludge volume reduction above 70% was achieved. However, more energy is required to achieve a significant volume reduction than sludge sanitization. Organic stabilization of the FS matter was expectedly not achieved with the MW treatment within the range of the test conditions in this study. The study was carried out in slum conditions, which are relatively similar to emergencies, so the results can be extrapolated to emergency settings. Further research is required to evaluate at a larger scale the performance and energy consumption of a customized MW unit under similar field conditions as tested in this study. With the large-scale unit, an economic analysis and comparison with other relevant technologies can be performed while the microbial indicators can be expanded to a wider specter of pathogens (e.g. viruses).

Acknowledgments

This research is funded by the Bill & Melinda Gates Foundation under the framework of SaniUP project (Stimulating Local Innovation on Sanitation for the Urban Poor in Sub-Saharan Africa and South-East Asia) (OPP1029019). The authors would like to thank Sanergy Kenya and the Asian Institute of Technology, Thailand for their valuable support during this study.

References

APHA, 2012. Standard Methods for the Examination of Water and Wastewater, 22nd ed. American Public Health Association, Washington DC, USA.

Banik, S., Bandyopadhyay, S., Ganguly, S., 2003. Bioeffects of microwave—a brief review. Bioresour. Technol. 87, 155-159.

Bastable, A., Lamb, J., 2012. Innovative designs and approaches in sanitation when responding to challenging and complex humanitarian contexts in urban areas. Waterlines 31, 67-82.

Belcher, D., Foutch, G.L., Smay, J., Archer, C., Buckley, C.A., 2015. Viscous heating effect on deactivation of helminth eggs in ventilated improved pit sludge. Water Sci. Technol. 72, 1119-1126.

Bennamoun, L., Arlabosse, P., Léonard, A., 2013. Review on fundamental aspect of application of drying process to wastewater sludge. Renew. Sust. Energ. Rev. 28, 29-43.

Border, B.G., Rice-Spearman, L., 1999. Microwaves in the laboratory: Effective decontamination. Clin. Lab. Sci. 12, 156-160.

Bresters, A.R., Coulomb, I., Matter, B., Saabye, A., Spinosa, L., Utvik, Å.Ø., 1997. Management Approaches and Experiences: Sludge Treatment and Disposal., Environmental Issues Series 7. European Environment Agency, Copenhagen, Denmark, p. 54.

Byamukama, D., Kansiime, F., Mach, R.L., Farnleitner, A.H., 2000. Determination of Escherichia coli Contamination with Chromocult Coliform Agar Showed a High Level of Discrimination Efficiency for Differing Fecal Pollution Levels in Tropical Waters of Kampala, Uganda. Appl. Environ. Microbiol. 66, 864-868.

Conder, G.A., Williams, J.F., 1983. The Microwave Oven: A Novel Means of Decontaminating Parasitological Specimens and Glassware. J. Parasitol. 69, 181-185.

Feachem, R.D., Bradley, D.J., Garelick, H., Mara, D.D, 1983. Sanitation and Disease Health Aspects of Excreta and Wastewater Management. World Bank Studies in Water Supply and Sanitation 3. The World Bank, Washington DC, USA, p. 501.

Fidjeland, J., Magri, M.E., Jönsson, H., Albihn, A., Vinnerås, B., 2013. The potential for self-sanitisation of faecal sludge by intrinsic ammonia. Water Res. 47, 6014-6023.

Flaga, A., 2005. Sludge drying, in: Plaza, E., Levlin, E. (Eds.), Proceedings of Polish-Swedish seminars. Integration and optimisation of urban sanitation systems., Cracow, Poland. Retrieved from http://www2.lwr.kth.se/Forskningsprojekt/Polishproject/index.asp?entry=13#Report13 .

Gan, Q., 2000. A case study of microwave processing of metal hydroxide sediment sludge from printed circuit board manufacturing wash water. Waste Manage. (Oxford) 20, 695-701.

Haque, K.E., 1999. Microwave energy for mineral treatment processes—a brief review. Int. J. Miner. Process. 57, 1-24.

Harvey, P.A., 2007. Excreta Disposal in Emergencies: a Field Manual. Water, Engineering and Development Centre (WEDC), Loughborough University, Leicestershire, UK.

Hong, S.M., Park, J.K., Lee, Y.O., 2004. Mechanisms of microwave irradiation involved in the destruction of fecal coliforms from biosolids. Water Res. 38, 1615-1625.

Hong, S.M., Park, J.K., Teeradej, N., Lee, Y.O., Cho, Y.K., Park, C.H., 2006. Pretreatment of Sludge with Microwaves for Pathogen Destruction and Improved Anaerobic Digestion Performance. Water Environ. Res. 78, 76-83.

Ingallinella, A.M., Sanguinetti, G., Koottatep, T., Montanger, A., Strauss, M., 2002. The challenge of faecal sludge management in urban areas - strategies, regulations and treatment options. Water Sci. Technol. 46, 285-294.

Jimenez, B., Austin, A., Cloete, E., Phasha, C., 2006. Using Ecosan sludge for crop production. Water Sci. Technol. 54, 169-177.

Katukiza, A.Y., Ronteltap, M., Niwagaba, C.B., Foppen, J.W.A., Kansiime, F., Lens, P.N.L., 2012. Sustainable sanitation technology options for urban slums. Biotechnol. Adv. 30, 964–978.

Kengne, I.M., Akoa, A., Soh, E.K., Tsama, V., Ngoutane, M.M., Dodane, P.H., Koné, D., 2008. Effects of faecal sludge application on growth characteristics and chemical composition of Echinochloa pyramidalis (Lam.) Hitch. and Chase and Cyperus papyrus L. Ecol. Eng. 34, 233-242.

Koné, D., Cofie, O., Zurbrügg, C., Gallizzi, K., Moser, D., Drescher, S., Strauss, M., 2007. Helminth eggs inactivation efficiency by faecal sludge dewatering and co-composting in tropical climates. Water Res. 41, 4397-4402.

Lamb, A.S., Siores, E., 2010. A Review of the Role of Microwaves in the Destruction of Pathogenic Bacteria, in: Anand, S.C., Kennedy, J.F., Miraftab, M., Rajendran, S. (Eds.), Medical and Healthcare Textiles. Woodhead Publishing, pp. 23-29.

Mawioo, P.M., Rweyemamu, A., Garcia, H.A., Hooijmans, C.M., Brdjanovic, D., 2016. Evaluation of a microwave based reactor for the treatment of blackwater sludge. Sci. Total Environ. 548–549, 72-81.

Menéndez, J.A., Domínguez, A., Inguanzo, M., Pis, J.J., 2005. Microwave-induced drying, pyrolysis and gasification (MWDPG) of sewage sludge: Vitrification of the solid residue. J. Anal. Appl. Pyrolysis 74, 406-412.

Menéndez, J.A., Inguanzo, M., Pis, J.J., 2002. Microwave-induced pyrolysis of sewage sludge. Water Res. 36, 3261-3264.

Moodley, P., Archer, C., Hawksworth, D., Leibach, L., 2008. Standard Methods for the Recovery and Estimation of Helminth Ova in Wastewater, Sludge, Compost and Urine Diversion Waste in South Africa. Water Research Commission, Pretoria, South Africa.

Mun, S., Cho, S.-H., Kim, T.-S., Oh, B.-T., Yoon, J., 2009. Inactivation of Ascaris eggs in soil by microwave treatment compared to UV and ozone treatment. Chemosphere 77, 285-290.

Pebsworth, P.A., Archer, C.E., Appleton, C.C., Huffman, M.A., 2012. Parasite Transmission Risk From Geophagic and Foraging Behavior in Chacma Baboons. Am. J. Primatol. 74, 940-947.

Remya, N., Lin, J.-G., 2011. Current status of microwave application in wastewater treatment—A review. Chem. Eng. J. 166, 797-813.

Richard, C., 2001. Excreta-related infections and the role of sanitation in the control of transmission, in: Lorna, F., Jamie, B. (Eds.), Water Quality: Guidelines, Standards & Health: Assessment of Risk and Risk Management for Water-Related Infectious Disease. IWA Publishing, London, UK., pp. 89-113.

Ronteltap, M., Dodane, P.-H., Bassan, M., 2014. Overview of Treatment Technologies, in: Strande L, Ronteltap M, Brdjanovic D (Eds.), Faecal Sludge Management - Systems Approach Implementation and Operation. IWA Publishing, London, UK, pp. 97-120.

Sangadkit, W., Rattanabumrung, O., Supanivatin, P., Thipayarat, A., 2012. Practical coliforms and Escherichia coli detection and enumeration for industrial food samples using low-cost digital microscopy. Procedia Eng. 32, 126-133.

Shamis, Y., Taube, A., Shramkov, Y., Mitik-Dineva, N., Vu, B., Ivanova, E.P., 2008. Development of a microwave treatment technique for bacterial decontamination of raw meat. International Journal of Food Engineering 4.

Tang, B., Yu, L., Huang, S., Luo, J., Zhuo, Y., 2010. Energy efficiency of pre-treating excess sewage sludge with microwave irradiation. Bioresour. Technol. 101, 5092-5097.

The Sphere Association, 2018. The Sphere Handbook: Humanitarian Charter and Minimum Standards in Humanitarian Response, 4rd ed. Practical Action Publishing, Rugby, United Kingdom, p. 406.

Thostenson, E.T., Chou, T.W., 1999. Microwave processing: fundamentals and applications. Compos. A: Appl. Sci. Manuf. 30, 1055-1071.

Tyagi, V.K., Lo, S.-L., 2012. Enhancement in mesophilic aerobic digestion of waste activated sludge by chemically assisted thermal pretreatment method. Biorcsour. Technol. 119, 105-113.

Tyagi, V.K., Lo, S.-L., 2013. Microwave irradiation: A sustainable way for sludge treatment and resource recovery. Renew. Sust. Energ. Rev. 18, 288-305.

Valero, A., Cejudo, M., García-Gimeno, R.M., 2014. Inactivation kinetics for Salmonella Enteritidis in potato omelet using microwave heating treatments. Food Control 43, 175-182.

WHO, 2006. Wastewater and Excreta use in Aquaculture, Guidelines for the safe use of wastewater, excreta and greywater. WHO, Geneva, Switzerland, p. 219.

Yu, Q., Lei, H., Li, Z., Li, H., Chen, K., Zhang, X., Liang, R., 2010. Physical and chemical properties of waste-activated sludge after microwave treatment. Water Res. 44, 2841-2849.

Chapter 7
Design, development and evaluation of a pilot-scale microwave based technology for sludge sanitization and drying

This chapter is based on:

Mawioo, P.M, Garcia, H.A., Hooijmans, C.M., Velkushanova, K., Simonič, M., Mijatović, I., Brdjanovic, D., 2017. A pilot-scale microwave technology based reactor for sludge drying and sanitization. Science of the Total Environment 601–602, 1437-1448.

Abstract

Large volumes of sludge are produced from onsite sanitation systems in densely populated areas (e.g. slums and emergency settlements) and wastewater treatment facilities that contain high amounts of pathogens. There is a need for technological options which can effectively treat the rapidly accumulating sludge under these conditions. This study explored a pilot-scale microwave (MW) based reactor as a possible alternative for rapid sludge treatment. The reactor performance was examined by conducting a series of batch tests using centrifuged waste activated sludge (C-WAS), non-centrifuged waste activated sludge (WAS), faecal sludge (FS), and septic tank sludge (SS). Four kilograms of each sludge type were subjected to MW treatment at a power of 3.4 kW for various time durations ranging from 30 to 240 minutes. During the treatment the temperature change, bacteria inactivation (*E. coli*, coliforms, *staphylococcus aureus*, and *enterococcus faecalis*) and sludge weight/volume reduction were measured. Calorific values (CV) of the dried sludge and the nutrient content (total nitrogen (TN) and total phosphorus (TP)) in both the dried sludge and the condensate were also determined. It was found that MW treatment was successful to achieve a complete bacterial inactivation and a sludge weight/volume reduction above 60%. Besides, the dried sludge and condensate had high energy ($\geq$ 16 MJ/kg) and nutrient contents (solids; TN $\geq$ 28 mg/g TS and TP $\geq$ 15 mg/g TS; condensate TN $\geq$ 49 mg/L TS and TP $\geq$ 0.2 mg/L), having the potential to be used as biofuel, soil conditioner, fertilizer, etc. The MW reactor can be applied for the rapid treatment of sludge in areas such as slums and emergency settlements.

7.1 Introduction

Various waste materials (e.g. sludge) are generated from onsite sanitation systems and wastewater treatment facilities. Onsite sanitation technologies, particularly portable toilets, pit latrines, septic tanks, among others, are commonly applied in densely populated areas (e.g. slums and emergency settlements). Therefore, these systems are intensively used requiring frequent emptying (Brdjanovic et al., 2015), which results in large quantities of fresh faecal sludge (FS) and septic tank sludge (SS). Conventional wastewater treatment plants (WWTP) are also sources of high sludge quantities which are usually produced during primary and biological treatment (Hong et al., 2004). The large volumes of sludge from both the onsite sanitation systems (e.g. FS and SS) and the conventional WWTPs (e.g. waste activated sludge) with high number of pathogens, and high portion of water and organic matter in it can lead to excessive handling and disposal costs, disease outbreaks, and offensive odor (Mawioo et al., 2016a; Mawioo et al., 2016b). In addition, vector attractions are major concerns during the final disposal of the sludge (Hong et al., 2004).

These concerns are particularly occurring in areas with rapid sludge generation, requiring responsive collection, transport, and treatment. Rapid sludge generation limits the application of traditional alternatives for sludge treatment and/or disposal. For example, the existing FS treatment options, e.g. composting, co-composting with organic solid waste, conventional drying, anaerobic co-digestion with organic solid waste, and co-treatment in wastewater treatment plants (Ingallinella et al., 2002; Katukiza et al., 2012; Ronteltap et al., 2014) are mostly suited for regular sanitation contexts, and they have limitations to their application in highly populated areas (Mawioo et al., 2016a; Mawioo et al., 2016b). Of main concern is the relatively low conversion rate of the processes involved in these technologies which usually have difficulties to match the high sludge production rates in such areas. Furthermore, the common sewage sludge treatment and/or disposal options e.g. landfill, agricultural application, ocean dumping, etc. (Menéndez et al., 2002; Lin et al., 2012) are facing increasing pressure due to lack of land space for landfills and the stricter regulations regarding pollution of the farmlands and water bodies. Other alternatives applicable for sewage sludge such as anaerobic digestion, composting, etc., which have relatively low conversation rates of the processes involved may not be viable where rapid sludge processing is required. These limitations demonstrate the need for development of technologies that are adapted to the specific field conditions and that should address the priority issues in sludge management with the aim to substantially reduce pathogen and sludge volume and, if desired, sludge stabilization. However, the reduction of the number of pathogenic organisms is priority, particularly in the densely populated conditions, due to the eminent risk of disease outbreaks. Moreover, because the densely populated conditions, e.g. slums and emergency settlements are often characterized by constraints in land space and time, it is required that the applied treatment technology is rapid and efficient, compact, and easy to install and deploy.

A promising technology to achieve these requirements is the use of a MW based technology (Mawioo et al., 2016a). The MW energy is a part of the electromagnetic spectrum with wavelengths (λ) ranging from 1 mm to 1 m and frequencies between 300 MHz (λ =1m) and 300 GHz (λ =1mm) (Haque, 1999; Tang et al., 2010; Remya and Lin, 2011). The MW technology is widely applied in heating applications, and its unique operational principle offers many advantages over the conventional heating (Haque, 1999; Thostenson and Chou, 1999). For instance, in contrast to the conventional thermal processes the MW energy offers benefits such as high heating rates, interior heating, energy saving, greater control of the heating process, and higher level of safety and automation, among others (Haque, 1999; Thostenson and Chou, 1999). Heating of a material by MWs results from the rotation of dipolar species and/or polarization of ionic species due to their interaction with the electromagnetic field (Haque, 1999). The molecular rotation and migration of ionic species causes friction, collisions, and disruption of hydrogen bonds within water; all of which result in the generation of heat (Venkatesh and Raghavan, 2004). The ability of a material to absorb MW energy and subsequently get heated is governed by its dissipation factor, which is the ratio of the dielectric loss factor to the dielectric constant of the material. The dielectric loss factor depicts the amount of input MW energy that is lost by being converted (dissipated) to heat within the material while dielectric constant depicts the ability of material to delay or retard MW energy as it passes through. Hence, materials with high dielectric loss factors are easily heated by MW energy (Haque, 1999). A low loss material can be heated indirectly by MW energy by blending with a high loss material (i.e. MW facilitator e.g. char). The MW first heats the facilitator which then heats the low loss material by conduction (Haque, 1999; Menéndez et al., 2002).

FS, SS, and WAS contains high amount of dipolar molecules such as water and organic complexes, which makes them good candidates for the MW dielectric heating. Preliminary laboratory studies have demonstrated the efficiency of MW technology regarding sludge sanitization and volume reduction (Mawioo et al., 2016a; Mawioo et al., 2016b). For instance, bacterial removal below the detection limit was reported when sewage sludge (Hong et al., 2004; Hong et al., 2006) and blackwater sludge (Mawioo et al., 2016b) were exposed to MW irradiation. In addition, a reduction below the detection limit of both *E.coli* and helminth (*Ascaris lumbricoides*) eggs was achieved by exposing fresh FS to MW irradiation (Mawioo et al., 2016a). Furthermore, a volume reduction above 70% was attained by the MW treatment of anaerobic sewage sludge (Menéndez et al., 2002), blackwater sludge (Mawioo et al., 2016b), and fresh FS (Mawioo et al., 2016a). The mechanisms associated with the sludge treatment by MW regarding pathogen and volume reduction include thermal (temperature) and the non-thermal (electromagnetic radiation) effects of the electromagnetic energy (Banik et al., 2003; Hong et al., 2004; Mawioo et al., 2016b). A distinguished aspect of the MW treatment is a combined effect of thermal and non-thermal action involved in the destruction of microorganisms. The thermal effect causes rapturing of microbial cells when water is rapidly heated to the boiling point by rotating dipole molecules under an oscillating electromagnetic field (Tang et al., 2010; Tyagi and Lo, 2013). Conversely, the non-thermal effects cause disintegration by the breakage of hydrogen bonds, which is attributed to the rapidly changing dipole orientation in the polarized side chains of the cell membrane macromolecules (Banik et al., 2003; Park et al., 2004; Tyagi and Lo, 2013; Serrano et al., 2016). Thermal effects of the electromagnetic energy are linked to the sludge volume reduction (sludge dehydration) during the MW treatment. The resulting high temperature causes the vaporization and eventually removal of the water contained in the sludge (Mawioo et al., 2016b). Notwithstanding its success in achieving sludge sanitization and volume reduction, the MW treatment did not attain organic stabilization of sludge; presumably, due to the relatively low maximum temperature attained in the treatment process, i.e. 127 °C (Mawioo et al., 2016b) and 134 °C (Mawioo et al., 2016a). The need for high temperature for the effective removal of organic matter was demonstrated when sludge was mixed with a better MW receptor (facilitator) and then irradiated to attain temperature of over 900 °C (Menéndez et al., 2002).

These studies demonstrate the potential of application of the MW technology to treatment of various types of sludge. They were conducted at a laboratory scale using relatively small quantities of sludge, but the findings serve as a solid base for further scaling up of the MW-based sludge treatment technology. Furthermore, the research so far has been focused mainly on sludge sanitization and volume reduction and has not assessed the potential value for the valorization of the end-products. It is therefore desirable to evaluate the MW-based sludge treatment technology at a larger scale with larger sludge quantities, while simulating closely real field conditions. In addition, it is important to assess the resource potential of the process end-products (i.e. dry sludge in terms of its CV and nutrient content, and condensate/product water in terms of nutrient content and quantity). When successfully developed and tested, the MW technology can provide a viable alternative for dealing with the complex task of sludge treatment and disposal, especially in the conditions where from high amounts are generated such as the WWTPs and densely populated areas (e.g. slums and emergency settings), and other similar conditions (e.g. public events and religious gatherings).

Basing on the successful outcomes of the studies discussed above, this study was planned with two main objectives. The first objective was to design and develop the microwave technology for sludge treatment at a larger (pilot) scale. The entire process involved a number of activities including the first step in which preliminary studies (Mawioo et al., 2016a; Mawioo et al., 2016b) were separately carried out using a domestic microwave to validate the technology using blackwater sludge (in Delft, The Netherlands) and faecal sludge (in Nairobi, Kenya). The outcomes of these studies are presented in Chapter 5 and 6, respectively. The second step involved design and development of a pilot microwave reactor unit by researchers at IHE Delft Institute for Water Education, Delft, the Netherlands and Fricke und Mallah Microwave Technology GmbH, Peine, Germany. The third step, which involved preliminary tests and improvement/modification of the pilot reactor unit was carried out by the researchers from IHE Delft Institute for Water Education, Delft, The Netherlands and Tehnobiro doo, in Maribor, Slovenia.

The second objective, which formed the core of this study was to evaluate the MW efficiency (on the basis of selected parameters) on the treatment of several types of sludge including WAS from WWTPs, C-WAS, SS, and fresh FS using the pilot-scale MW reactor unit. In addition, the added value of the treatment end-products was evaluated. The study focused on four aspects, namely: reduction of selected pathogens, sludge volume reduction, organic matter reduction, and assessment of the value of treatment end-products. *E. coli*, coliforms, *staphylococcus aureus* and *enterococcus faecalis* bacteria were used as pathogenic indicator microorganisms, while the sludge weight was used to estimate the volume reduction. CV, TN and TP content were used to assess the value of the treatment end products, while organic stabilization of the FS was estimated using the volatile to total solids ratio (VS/TS) comparison.

7.2 The concept

The idea to apply the microwave technology for sludge treatment was conceived as part of a larger innovative concept for the faecal sludge management in complex conditions, e.g. the disaster situations. A shown in Figure 7-1, the main components of the concept include containment facilities (toilets) for the collection/storage of the faecal sludge, a dewatering unit (based on centrifugation) to reduce the sludge water content, and an innovative membrane bioreactor and microwave reactor units for the respective treatment of the liquid (concentrate) and the solids (sludge cake) that result from the centrifugation process.

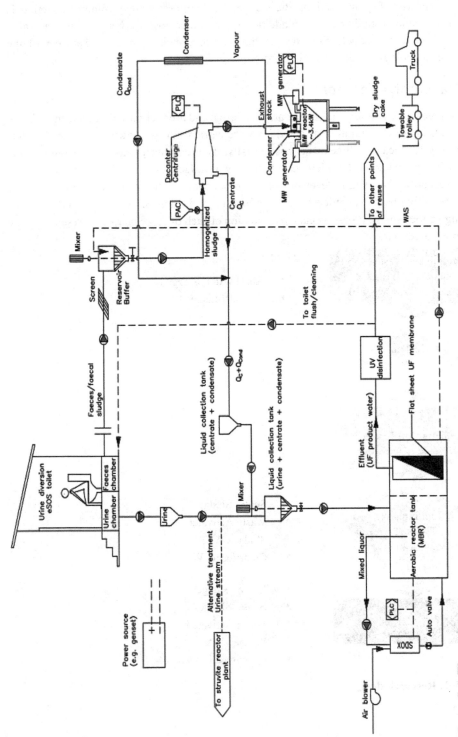

Figure 7-1. The concept

The microwave unit, which is the focus of this study, is aimed at pathogen destruction as a priority to ensure safety in handling the sludge. Furthermore, sludge volume reduction is expected (to reduce handling costs) in addition to any other possible benefits such as value addition of the end products from the treatment. The principles and mechanisms of the microwave technology are discussed in Chapter 4.

7.3 Research approach

The overall goal of this study was particularly the design and production a pilot scale microwave reactor unit, in which a number of activities were carried out as shown in Figure 7-2. The research was conducted in three main phases; firstly: the preliminary tests at a laboratory scale using a conventional domestic microwave to treat two sludge types (i.e. blackwater and faecal sludge), secondly: the design, development and production of the microwave pilot unit, and thirdly: the preliminary evaluation tests, modifications and improvements of the pilot reactor unit using various sludge types. A detailed discussion of each of those phases is presented in the following subsections.

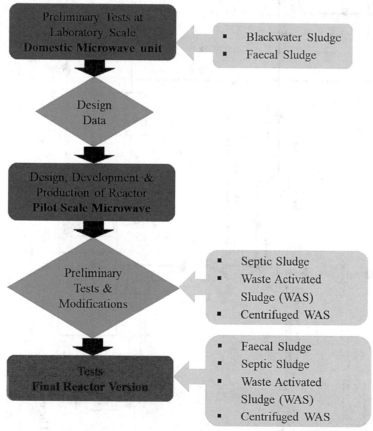

Figure 7-2. Research design

7.4 Validation, design, development and production

7.4.1 Validation tests

The initial stages involved validation tests to assess the applicability of the microwave irradiation technology for sludge treatment and acquire preliminary data for the design of the pilot unit. A domestic microwave unit was used for the validation tests in which two sludge types were used. The first batch of tests was conducted using blackwater sludge obtained by centrifuging raw blackwater generated from low flush toilets located in a demonstration plant in Sneek, The Netherlands, while the second batch of tests was conducted using faecal sludge generated from intensively used UDDT toilets located in the slums of Nairobi, Kenya. The results of these two sets of study are presented in Chapter 5 and 6, respectively.

7.4.2 Design and production of the pilot unit

Based on the results previously obtained in the preliminary laboratory scale tests (Chapter 5 and 6), a pilot-scale microwave reactor unit was designed by researchers at the IHE Delft Institute for Water Education (Delft, The Netherlands), and then developed and manufactured in collaboration with Fricke und Mallah Microwave Technology GmbH (Peine, Germany) with additional modifications carried out by Tehnobiro d.o.o. (Maribor, Slovenia). A design based on a conical shape mixed reactor intended for the treatment of sludge in either continuous or batch mode with a total installed power capacity of 40kW (i.e. comprising four microwave generators each with 10kW power output) was originally planned Figure 7-3. However, financial limitations allowed the purchase of four microwaves generators with a total installed power of 3.4 kW and an on-the-shelf cavity of a much larger volume than originally planned Figure 7-3.

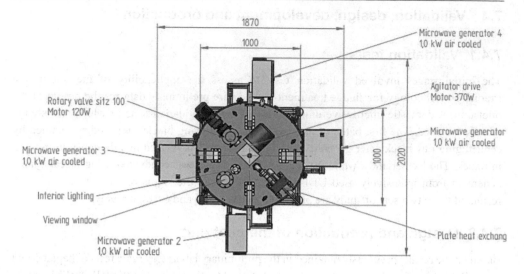

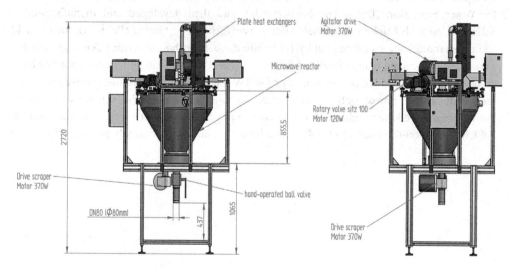

Figure 7-3. Model design of the microwave reactor unit (courtesy of Fricke und Mallah Microwave Technology GmbH)

Further, a series of preliminary tests, improvements and modifications were carried out to arrive at the final reactor version as discussed in the subsequent sections.

7.5 Preliminary experimental tests and improvements

As mentioned in Section 7.4.2 a series of preliminary tests along with modifications were carried out to improve the prototype microwave reactor unit. In this process various versions of the unit including version 1 (i.e. the original version), 2 and 3 (i.e. final version) were realized as discussed below.

7.5.1 The pilot-scale microwave reactor unit version 1

The main features of the original unit, i.e. version 1 shown in Figure 7-4 comprised a 200 L stainless steel cavity (i.e. the applicator) for holding the sludge during irradiation, four air-cooled microwave generators, a condenser unit, an infrared temperature sensor, and a control unit.

Figure 7-4. The microwave reactor unit version 1 (Photo: D. Brdjanovic)

The four microwave generators with a total installed power of 3.4 kW (i.e. 850W each) at 2,450 MHz were installed on the top cover of the reactor. The condenser unit was also installed on the reactor cover for ease of vapour collection and odour control. The optris®CSmicro infrared temperature sensor (Optris GmbH, Germany) continuously measured the inside cavity temperature during the treatment. All components of the control unit were integrated and supported by a control software (LabVIEW 8.0, National Instruments Corporation, USA) installed in a control computer (Dell laptop, Dell Inc., USA) that was used to the operation control and data logging. As a precautionary measure, magnetic contactors were also installed between the reactor tank and the top cover to avoid accidental switch-on of the magnetrons whenever the cover is open. The reactor unit was evaluated and subsequently underwent a series of modifications based on the preliminary test results. The entire process of tests and modifications was carried out at the workshop facility of Tehnobiro d.o.o. in Maribor, Slovenia during the period of 60 days (January to February 2016). Prior to the practical experimental tests with sludge samples, a simulation study was carried out that was aimed to theoretically visualize the power distribution in the system (Figure 7-5).

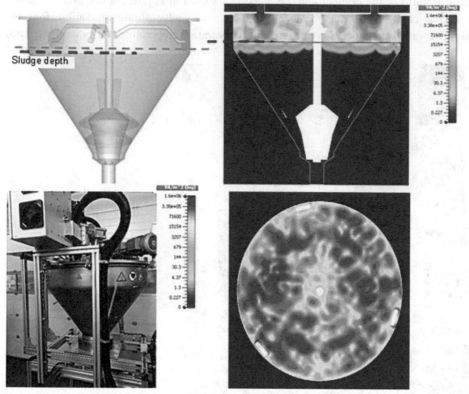

Figure 7-5. A MW model showing the power intensity distribution expressed as VA/m² (W/m²) in the sludge during a microwave treatment trial (courtesy of Fricke und Mallah Microwave Technology GmbH)

The results of the power distribution model, which is an indication of the expected resultant temperature progression are shown in Figure 7-5. The results demonstrated that the microwave irradiation was most efficient within the top part of the cavity (15–20 cm), in which the red colour illustrates the area with the highest power intensity (hottest zones) and blue the area with the lowest power intensity (i.e. the coldest zones).

Furthermore, practical tests which were mainly focussed on the temperature progression in the system were carried out on the unit version 1 using four sludge types namely; primary sludge, waste activated sludge, centrifuged waste activated sludge, and septic tank sludge all collected from the Čistilna Naprava WWTP, located in Ptuj, Slovenia.

The primary sludge and the waste activated sludge were obtained from the bottom parts of the primary and secondary settling tanks, respectively. The centrifuged waste activated sludge was obtained at the discharge pipe of the centrifuge unit while septic tank sludge was collected from the trucks as they discharged into the WWTP. The samples were collected in plastic containers and transported to the test facility where they were stored before running the experiments. For each sample type a thorough mixing was carried out on the bulk sample from which smaller test samples in batches of 100 litres were obtained using a graduated bucket. The samples were then separately transferred into the microwave reactor cavity via the rotary feeder and exposed to the microwave irradiation at 3.4 kW for variable time periods between 0 – 360 minutes. To determine the effect of the microwave treatment, samples were collected every 30 minutes during the treatment for each sample type and analysed for various parameters. The final volume was also measured to assess volume reduction (i.e. extend of sludge drying) by the microwave heating. Furthermore, the vapour that was produced from the heated sludge was cooled down by the condenser unit and the generated condensate was collected in a plastic container for analysis. Moreover, sludge temperature was measured on the surface of the reactor cavity and in the collected sludge sample every 30 minutes using a thermal camera (FLIR TG 165, FLIR Systems Inc., US).

The temperature progression results shown in Figure 7-6 confirmed that the principle works, but it took too long to heat up and dry the sludge in such a system. For instance, it took approximately 360 minutes to raise the test sludge sample temperature to approximately 66°C. In addition, the results of the model study (shown in Figure 7-5) that the MW irradiation was most efficient within the top part of the cavity (15-20 cm) were confirmed by the practical observations. It was also observed that the test samples displayed relatively slight differences in the temperature evolution rates, which can be attributed to their specific composition characteristics.

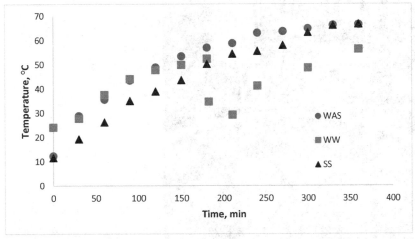

Figure 7-6. Effect of exposure to microwave irradiation on temperature in the 100 L test samples with reactor version 1

The observations on the reactor unit version 1 demonstrate that despite the MW capability to cause volumetric heating of material, there is an optimal depth by which the waves can penetrate a specific material. The penetration depths for specific materials should thus be considered when designing MW reactors. Basing on the test results, it was necessary to improve conditions in the system to ensure increased exposure and contact between the test sample and the microwaves. Therefore, further modifications performed on the reactor to achieve the reactor version 2 as described below.

7.5.2 The microwave reactor unit version 2

The design features of the rector version 2 were generally similar to that of version 1 described in section 7.5.1 above. The only difference was both the recirculation pump and an additional agitator that were introduced in the reactor version 2 for sludge recirculation to enhance its mixing and exposure to the microwaves at the top of the reactor (Figure 7-7).

Figure 7-7. Reactor Version 2 with an installed recirculation pump (top) and a mechanical mixer (bottom)

The recirculation pump worked only for the liquid samples, while the mechanical mixer aided in mixing the samples that were solid in nature e.g. centrifuged sludge and the liquid samples as they turned more solid along the drying process. Version 2 was further modified by introducing a rotational tray at about 20 cm from the top of the cavity. In addition, the amount of sludge sample introduced to the reactor was reduced from 100 kg to approximately 10 kg.

The bottom of the cavity was used as a storage of the sludge scraped off the tray still with the possibility to be recycled as long as the recycle pump could handle that sludge. Experimental tests were performed on the reactor with sludge similar to that used in the version 1. Based on the experience with the first two versions of the prototype it was decided to further modify the reactor design to optimise its performance. Therefore, further modifications targeting both the reactor volume and the sample size were carried out to achieve the final reactor unit version 3. Extensive evaluation tests were conducted on the reactor version 3 whose features and test results are presented in section 7.6 below.

7.6 Evaluation of pilot-scale microwave reactor unit version 3

Basing on a series of modifications and improvements as discussed above, the final reactor version 3 was developed and tested. The main objective of this study was to evaluate the MW efficiency (on the basis of selected parameters) on the treatment of several types of sludge including WAS from WWTPs, C-WAS, SS, and fresh FS. The research was conducted at pilot-scale, and the added value of the treatment end-products was evaluated. The study focused on four aspects, namely: reduction of selected pathogens, sludge volume reduction, organic matter reduction, and assessment of the value of treatment end-products. *E. coli*, coliforms, *staphylococcus aureus* and *enterococcus faecalis* bacteria were used as pathogenic indicator microorganisms, while the sludge weight was used to estimate the volume reduction. CV, TN and TP content were used to assess the value of the treatment end products, while organic stabilization of the FS was estimated using the volatile to total solids ratio (VS/TS) comparison.

7.7 Materials and methods

7.7.1 Research design

This study was performed using four different sludges including WAS, C-WAS, SS, and FS. The differences of properties between the tested sludges allowed to assess the capability of the MW-based technology to treat each type of sludge with a variability that can be expected in real life full-scale field applications. The study was divided in four phases with each phase involving one type of sludge, starting with WAS and C-WAS (less demanding in terms of preliminary experimental handling). The next phase involved the treatment of SS, concluding with the treatment of fresh FS which was considered as the most challenging sludge sample to treat. Each time a 4 kg sludge sample was exposed to irradiation provided by MW generators (total installed power 3.4 kW) for a certain time period. Maximum contact time necessary to attain complete drying for each of the sludge types was pre-determined on the basis of preliminary trials. Parameters such as temperature, number of bacteria (i.e. *E. coli*, coliforms, *staphylococcus aureus*, and *enterococcus faecalis*), weight/volume, and VS/TS ratio were identified to evaluate the MW unit's performance. These parameters were measured in both the raw (untreated) and treated sludge samples, and the differences were used to determine the reactor performance in time. The specific energy consumptions of the sludges at the various contact times were also calculated using the electrical energy supplied to the MW unit and the resultant loss in sludge weight/volume. Furthermore, to assess the value of the treatment end-products, additional parameters were measured for the treated sludge and the generated condensate at specific times during the experiment. These include COD, CV, TN, TP, and bacteriological indicators. All experiments were performed in triplicates. The experimental data was processed using Microsoft Excel software. In each case, the triplicate data for each set of contact time was combined by computing their mean values and the respective standard deviations and standard errors. The mean values were then presented in graphs with their respective standard error bars shown.

7.7.2 Experimental apparatus

The plot-scale MW reactor used in this study is a product of a serials of activities ranging from preliminary laboratory investigations, design, development preliminary tests and improvements as discussed in section 7.4 and 7.5 above. Based on the experience with the first two versions of the prototype, it was decided to replace the conical cavity/reactor with a flat bottomed and relatively shallow unit without changing the arrangement on the top cover. Mixing and recycling was omitted in this unit, and the working capacity of the reactor was kept to approximately maximum of 30 kg sludge. However, it was decided to bring down the actual sample in the batch type of experimentation to 4 kg to better match the installed power capacity and decrease the exposure time. This third (final) version of the prototype was in fact used in this study (see Figure 7-8).

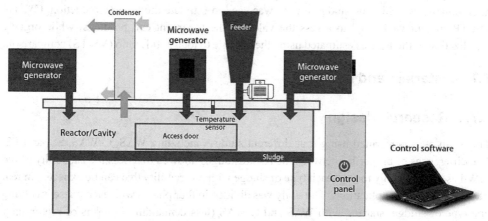

Figure 7-8. Schematic diagram of the microwave reactor unit

The reactor unit (i.e. the final version) comprised a 196 L stainless steel cavity (i.e. the applicator) for holding the sludge during irradiation. The cavity was cylindrical in shape with a diameter of approximately 100 centimeters and a depth of 25 centimeters. Other features included four air-cooled MW generators, a condenser unit, an infrared temperature sensor, and a control unit. The four MW generators each with 850 W at 2,450 MHz were installed on the top cover of the reactor. The condenser unit was also installed on the reactor cover for ease of vapour collection and odour control. The optris®CSmicro infrared temperature sensor (Optris GmbH, Germany) continuously measured the inside cavity temperature during the treatment. However, due to the instability noticed with the measurements by the sensor, a thermal camera (FLIR TG 165, FLIR Systems Inc., USA) was also used to measure the final (maximum) temperature of the irradiated sludge at the end of each treatment cycle. The temperature measured by the thermal camera was more reliable and thus was adopted for reporting in this study. Furthermore, the reactor was covered with a 10 cm fibreglass wool insulation layer with aluminium foil to reduce temperature losses to the surrounding environment. The ambient temperature (inside the industrial hall) was around 10°C during the experimental period. As an extra measure of caution, a MW leak detector (Tenma 72-10212, Tenma, China) was used to check any MW leakages around the reactor. All components of the control unit were integrated and supported by a control software (LabVIEW 8.0, National Instruments Corporation, USA), which was installed in a control computer (Dell laptop, Dell Inc., USA) that was used to control the operation and log the data. A power meter was used to measure the electrical energy supplied to the MW units, which together with the sludge weight/volume reduction was used to estimate the specific energy consumption at the various contact times.

The MW reactor was located at the workshop facility of Tehnobiro d.o.o. in Maribor, Slovenia where the entire experimental work was conducted during the period of 150 days (March to July 2016).

7.7.3 Sludge samples

The C-WAS, WAS and SS samples were collected from a municipal WWTP located in Ptuj, Slovenia. The WAS sample was collected from the WAS discharge line, while the C-WAS was obtained at the discharge point of the dehydrated WAS, which was thickened by the means of a centrifuge with the addition of a poly-electrolyte. The SS sample was collected from the trucks (while discharging at the WWTP) which collected the sludge from the septic tanks of the houses in the areas not covered by the sewerage. The fresh FS was obtained from urine diverting dry toilets (UDDTs) located at the Sustainability Park Istra (Gračišče, Slovenia). In each case, the samples were collected and transported in closed plastic containers to the research facility in the industrial zone of Maribor where they were stored at 4 °C prior to the experiments and analysis within 48 hours.

7.7.4 Sample preparation and treatment

The sludge samples were treated using the MW apparatus. The samples were thoroughly mixed and weighted into batches of 4 kg and placed in plastic containers. They were then transferred into the MW reactor unit's cavity (i.e. applicator), evenly spread to attain a thickness of approximately 0.5 centimeter, and then exposed to the MW irradiation at 3.4 kW for variable time. Four batches each for the C-WAS and FS samples were separately exposed for the respective maximum periods of 30, 60, 90, and 120 minutes, while five batches each for the SS and WAS samples were separately exposed for the respective maximum periods of 30, 60, 90, 120, and 240 minutes. As a consequence of the MW treatment, vapor was produced from the heated sludge. The vapor was cooled down by a condenser and the generated condensate was collected in a plastic container separately for each test. After the MW treatment, the cavity door was opened and the sludge sample temperature was immediately measured. The irradiated samples (Figure 7-9) were then cooled down and later analyzed for various parameters as described in the following section.

Figure 7-9. Irradiated samples; a) - Centrifuged waste activated sludge (C-WAS), b) - Faecal sludge (FS), c) - Non-centrifuged waste activated sludge (WAS), d) - Septic tank sludge (SS)

Various operational challenges were experienced at different stages of the treatment process. For instance, the loading process required spreading the sludge samples to achieve a relatively even thickness which consumed some time. This was particularly noticeable with the solid samples e.g. FS and C-WAS. It also took time to wait for the sludge temperature to cool down to a safe level for unloading, especially when samples were irradiated at longer contact time periods. Furthermore, dried samples got stuck on the cavity thus taking some time to scrape them off.

7.7.5 Analytical procedures

Several physical, chemical, and microbial parameters were measured in the sludge before and after the MW treatment and in the condensate. The sampling framework is shown in Table 7-1.

Table 7-1. Sampling framework

Parameter/Media	WAS	C-WAS	SS	FS	Concentrate[1]
Temperature (°C)	•	•	•	•	
Weight-volume reduction ratio	•	•	•	•	
Total Solids (mg TS)	•	•	•	•	
Volatile Solids (mg VS)	•	•	•	•	
Calorific value (kJ/kg)	•	•	•	•	
Organic matter (mg COD/g TS)	•	•	•	•	••
Total Nitrogen (mg N/g TS)[2]	•	•	•	•	••
Total Phosphorus (mg P/ g TS)[2]	•	•	•	•	••
E. coli (CFU/g TS)	•	•	•	•	••
Total coliforms (CFU/g TS)	•	•	•	•	••
Staphylococcus aureus (CFU/g TS)	•	•	•	•	••
Enterococcus faecalis (CFU/g TS)	•	•	•	•	••

[1] Condensate which is generated by treatment of each sludge sample.

[2] TN and TP concentrations were measured only after the MW treatment.

• Just before irradiation and immediately after the irradiation exposure

•• At the end of the test, taken from the total collected condensate during the test

Temperature measurement

The sample temperature was measured just before the exposure to microwaves. At the end of the each batch exposure, the cavity door was opened and the final (maximum) sample temperature was immediately measured in the reactor by a thermal camera (FLIR TG 165, FLIR Systems Inc., USA). The built in infrared temperature sensor (optris®CSmicro, Optris GmbH, Germany) located in the top plate of the cavity was continuously measuring the temperature, it was not accurate as it recorded the temperature of air close to the top of the reactor and not in the sludge sample.

Weight/volume reduction calculation

The initial weight of the sample was determined using a weighing scale. After each treatment, the sample was cooled to room temperature and its weight was measured again. The volume reduction was then determined from the difference between the two measurements. Based on the maximum sludge temperature attained during MW treatment (i.e. $\leq 102\ ^{\circ}C$), the weight reduction could mainly be attributed to the water evaporated from the heated sludge. Thus, considering the density of water, the weight reduction was assumed to be equivalent to the sludge volume reduction.

TS and VS measurements

TS and VS content of the sludge were determined according to the gravimetric methods (SM-2540D and SM-2540E, respectively, as described in APHA (2012).

Calorific value measurement

The gross CV of the treated/dried sludge was measured in a bomb calorimeter (IKA-Kalorimeter C 400 adiabatisch IKA®-Werke GmbH & Co. KG, Staufen German) based on the ISO 1928:2009 Standard.

Total COD measurement

Samples for COD measurement in the sludge and the condensate were prepared by diluting a known amount of sample in distilled water. The COD measurement was then conducted according to the open reflux method (SM 5220 B) (APHA, 2012). The values were expressed in mg COD per g TS (mg COD/g TS) or in mg COD per g liter (mg COD/L) for the condensate.

TN and TP measurement

TN in the solid and liquid samples was measured using Dumas and Kjeldahl methods, respectively (Etheridge et al., 1998). TP in the solid and liquid samples were measured using the ascorbic acid method with acid persulfate digestion.

Microbiological analyses

The detection of *E. coli*, coliforms, *staphylococcus aureus*, and *enterococcus faecalis* was done using the surface plate technique. Chromocult coliform agar was used for both *E.coli* and coliforms (Chromocult; Merck, Darmstadt, Germany) (Byamukama et al., 2000). A step by step procedure as explained in Mawioo et al. (2016b) was applied to prepare and incubate the samples for the *E. coli* and coliforms analysis. Dark blue to violet colonies were classified as *E. coli* while red to pink colonies were identified as coliforms (Byamukama et al., 2000; Sangadkit et al., 2012). *Staphylococcus aureus* and *enterococcus faecalis* were grown using Baird-Parker RPF agar, and Slanetz and Bartley agar media, respectively (bioMérieux SA, Marcy l'Etoile, France). The preparation involved spreading 0.1 mL of the sample or its dilutions on the surface of the agar plate. The inoculated plates for *enterococcus faecalis* were incubated at 37 °C for 48 hours and the colonies were identified as red, brown or pink. For the *staphylococcus aureus*, the inoculated plates were incubated at 37 °C for 24 hours but prolonged for an additional 24 hours if no characteristic colonies were formed. Their colonies were identified as grey-black. The average number of colonies were used to calculate the viable-cell concentrations in the samples and expressed in either CFU/g TS or CFU/L of the test sample.

7.8 Results and discussion

7.8.1 Characteristics of the sludge samples

The sludge characteristics are presented in Table 7-2. FS exhibited the highest dry matter content followed by C-WAS, SS, and WAS, respectively. FS, C-WAS and WAS had a relatively high organic matter content (i.e. VS/TS over 0.70) compared to the SS with a VS/TS of 0.55.

Table 7-2. Characteristics of the raw sludge

Parameter	Sludge type			
	C-WAS	FS	WAS	SS
Water content (%)	85	77	98	92
Total solids (TS, %)	15	23	2	8
VS/TS	0.76	0.88	0.76	0.55
TCOD (mg O$_2$/g TS)	1.5	1.4	5.6	2.7
E. coli (CFU/g TS)	Not detected	4.23×10^8	2.41×10^7	1.03×10^5
Coliforms	9.98×10^6	3.58×10^7	5.56×10^8	3.42×10^6
Staphylococcus aureus	3.00×10^7	1.51×10^8	9.74×10^7	1.11×10^5
Enterococcus faecalis	2.41×10^7	4.55×10^8	4.87×10^7	1.46×10^5

Furthermore, all measured pathogen indicators were detected in all sludges except the *E.coli* that was not detected in the C-WAS. As expected, the fresh FS appeared more polluted as it exhibited the highest pathogenic loads than all other sludges. The content of the various parameters as measured in the sludges were within the expected range. For instance, the values for TS and VS/TS in the FS were approximately 23% and 0.88, respectively, which is comparable to those reported in previous studies. Mawioo et al. (2016b) reported 26% and 0.92, while Rose et al. (2015) reported 25% and 0.84-0.93 for the TS and VS/TS ratio, respectively. Variations in the faecal composition can be expected due to diverse dietary intake of food and fluid (Rose et al., 2015). The TS (2%) and VS/TS ratio (0.76) values in the WAS were also comparable to those previously reported in other studies. For instance, Kim et al. (2003) reported a TS of 3.8% and VS/TS ratio of 0.68 for WAS while Tang et al. (2017) reported a TS of 2.1%. However, the TS value was relatively higher than the typical range (0.7 – 1.2%) (Takács and Ekama, 2008), which can be attributed to the fact that the sample was collected from the pipe delivering sludge to the centrifuges rather than in the clarifier. The TS content of the C-WAS (15%) was relatively lower than the typical 22-36% achieved using centrifuge, belt-filter press, and filter press dewatering (Metcalf and Eddy, 2003). Despite a high TS content (8%), SS had the lowest organic matter fraction with a VS/TS ratio of 0.55, which can be attributed to the high content of grit (observed) and the degradation of organic matter due to long retention time in the septic tanks.

7.8.2 Temperature evolution during the tests

Figure 7-10 shows the temperature profiles for the respective sludge samples when exposed to MW irradiation. The temperature response of the samples to MW treatment appeared to vary among the sludge types. Generally, the temperature increment rate was relatively higher in the sample with low water content (i.e. C-WAS and FS), with both achieving a maximum of approximately 102 °C. Conversely, WAS and SS, which had a relatively high water content, exhibited a comparatively low increment rate to achieve a maximum temperature of approximately 96 °C. However, it can also be seen that C-WAS portrayed a higher rate of temperature evolution than the FS that had a lower water content. A similar trend in the temperature evolution rates was observed for the WAS in comparison to the SS.

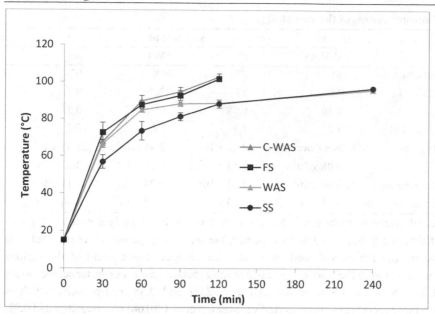

Figure 7-10. Temperature evolution in the sludge samples during the MW treatment

Furthermore, as shown in Figure 7-10, three distinctive temperature evolution phases were portrayed. These showed an initial rapid rise in sludge temperature that was followed by a fairly constant but minimal temperature rise, and finally again a fairly rapid temperature rise.

The variation in the temperature increment rates among the sludge samples evaluated can be attributed to their properties (Table 7-2), e.g. the water content, organic matter content, viscosity, etc. For instance, the samples with a higher water content (i.e. WAS and SS) demonstrated a relatively lower temperature increment rate. Water is a good dielectric material, but has a high thermal capacity, so more heat energy was initially consumed with a relatively smaller temperature increment in the samples with a higher water content compared to those with a lower water content (C-WAS and FS) (Mawioo et al., 2016b). This observation agrees with the previous studies in which different quantities and types of sludge with varying water contents were heated by microwaves (Tang et al., 2010; Mawioo et al., 2016a; Mawioo et al., 2016b). In addition, the different levels of organic matter among the sludges might have influenced their variation in temperature evolution. For example, WAS had a higher water content (98%) but still exhibited a higher temperature evolution rate than SS (92%), possibly due to its higher organic matter content i.e. VS/TS ratio of 0.76 compared to 0.55 in the SS. Carbohydrates, proteins and lipids, which are the primary organic components found in sludge (Tyagi and Lo, 2013) have high loss factors, hence their constituent amounts will definitely influence the sludge response to MW heating. Furthermore, the dry matter of the SS had a relatively high sand (quartz) content which has poor dielectric properties and is largely transparent to microwaves.

Conversely, the FS, which had a lower water but higher organic matter content than the C-WAS, exhibited a comparatively low temperature increase rate, which can be the effect of the viscosity. The FS sample was stickier, which can interfere with the movement of molecules and consequently the MW effect in the sludge. The MW irradiation effect is influenced by the friction of molecules in the medium, its change to heat energy, and the synergistic effect during irradiation through convection (Hong et al., 2004). Therefore, stickiness (viscosity) of a medium can influence temperature evolution during MW heating. However, the FS viscosity can be reduced by applying selected cover (odor control) materials such as char, etc., in the UDDTs so as to mix with FS in situ or by blending those materials with FS prior to the MW treatment. Besides odor control and viscosity improvement, char is a good MW receptor (has high loss factor) material that can enhance the MW irradiation process.

Despite attributing the differences in temperature evolution rates to the sludge properties, it was not possible to explicitly point out their specific range of influence. This can be achieved by conducting specific tests to determine the interactions of the various sludge properties with MW radiation, which was not in the scope of this study.

Furthermore, the temperature evolutions phases depicted in Figure 7-10, are similar to observations reported in the previous studies, in which the three stages were classified in reference to the sludge drying phases, namely the preliminary, essential (major), and final drying phases (Flaga, 2005; Bennamoun et al., 2013; Bennamoun et al., 2014; Bennamoun et al., 2016; Mawioo et al., 2016b). Those phases were associated with the changes (e.g. heating and evaporation) occurring during the sludge drying period. For instance, the rapid temperature rise during the preliminary drying phase was linked to the initially high concentration of dipolar molecules (e.g. water, proteins, etc.) in the wet sludge that interacted with the microwaves to generate high amounts of heat (Yu et al., 2010; Chen et al., 2014; Mawioo et al., 2016b). On the other hand, the fairly constant but minimal temperature rise during the essential drying phase was associated with the fact that majority of the heat goes to the vaporization of the unbound water from the surface of the sludge particles while being constantly replaced from the inside (Flaga, 2005; Mawioo et al., 2016b). During the final phase, the rapid temperature rise was attributed to the more rapid evaporation of water from the surface of the sludge particles than it is replaced from the inside of the particle (Mawioo et al., 2016b).

It is also worth to mention that temperature measurements in this case was a challenge as it required first to switch off the MW generators (for operator safety) and then open the cavity door to facilitate the measurements using the thermal camera. Although this was done in a quick succession, the measured final (maximum) sludge temperature levels might probably be relatively lower than those actually attained as some heat loss could be expected, especially when the cavity door was opened. Such limitation can be addressed by installing an inbuilt temperature probe that is extended into the sample to replace the infrared sensor which was found to be unreliable as it recorded the temperature of air close to the top of the reactor and not in the sample.

7.8.3 Pathogen reduction

Figure 7-11 shows the reduction profiles for the pathogens (selected indicator microorganisms) as function of the input MW energy for each type of sludge. The MW treatment was able to achieve complete destruction (below detection limit, i.e. <500 CFU/g TS or approximately 5.90 log removal value (LRV)) of pathogens of interest in the sludge at the conditions evaluated in this work. In all samples, the destruction of pathogens increased with increasing MW energy. For instance, in all samples when the sludge was exposed at 1.7 kWh, an over 3 LRV was achieved for the coliforms and *enterococcus faecalis* bacteria. However, when the exposure to MW energy was raised to 3.4 kWh, a reduction beyond the detection limit was achieved.

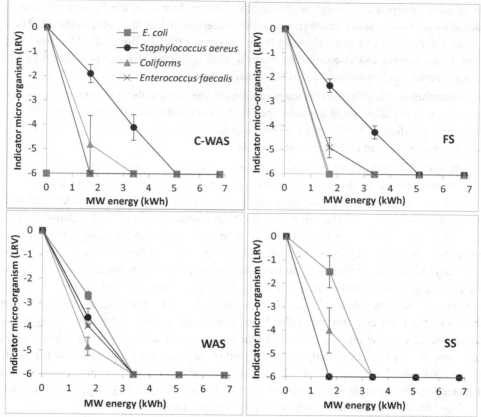

Figure 7-11. Effect of MW energy on reduction of indicator microorganisms in different sludges

Despite achieving a reduction below the detection limit for all selected pathogen indicator organisms in all the evaluated types of sludge, some disparities were observed concerning their die-off rate (Figure 7-11). For instance, in the C-WAS and FS samples a relatively slower reduction was observed of *staphylococcus aureus* (e.g. 2 LRV at 1.7 kWh) compared to both the coliforms and *enterococcus faecalis* (e.g. 4 LRV at 1.7 kWh). In addition, a reduction beyond the detection limit for the coliforms and *enterococcus faecalis* was observed at 3.4 kWh compared to the 5.1 kWh needed for *staphylococcus aureus*. Furthermore, a higher reduction rate for the *staphylococcus aureus* was observed in the samples with a higher initial water content (WAS and SS) compared to the samples with a lower initial water content (C-WAS and FS).

Moreover, the four pathogen indicator bacteria were measured in the product water/condensate, but none was detected.

The increment of the pathogen destruction with the increasing MW energy can be attributed to the resulting raise in both the thermal (temperature) effect and the electromagnetic radiation intensity (non-thermal effect), both of which are the mechanisms by which the microorganisms are destroyed during the MW treatment. Thermal effect is known to be the main mechanism by which the destruction occurs. Thermal effect causes rapturing of the microbial cells when water is rapidly heated to the boiling point by the rotating dipole molecules under an oscillating electromagnetic field (Tang et al., 2010; Tyagi and Lo, 2013). A temperature of approximately 70 °C has been reported as essential for the bacterial destruction (Hong et al., 2004; Valero et al., 2014). Microbial cells are also destructed by the non-thermal (electromagnetic radiation) effects by disruption of the hydrogen bonds that results when molecules or ionic species in the cell continually orient themselves in the direction of the changing electromagnetic field (Banik et al., 2003; Park et al., 2004; Tyagi and Lo, 2013; Serrano et al., 2016). Nevertheless, it is still difficult to distinguish between thermal and non-thermal effects in the relative contribution to pathogen kill.

The disparity concerning the rate of reduction, especially of the *staphylococcus aureus* between the solids and the liquid samples can be linked to the sludge properties, e.g. dry matter content and viscosity, which might have affected the uniformity of the distribution of heat and electromagnetic radiation in the sample. A likely influence of the media properties was demonstrated in Walker and Harmon (1966) who observed varied thermal destruction rates of *staphylococcus aureus* in skim milk, whole milk, cheddar cheese whey media, and a phosphate buffer.

Furthermore, the absence of the pathogen indicator bacteria in the product water/condensate can be attributed to the higher temperature necessary for water vaporization (i.e. 100 °C) compared to the essential 70 °C for bacteria destruction.

Generally, the results obtained here demonstrate that MW can offer an effective solution to sludge sanitization. The technology was capable to achieve complete reduction for all the pathogen indicator bacteria tested in the different sludge types. The results are in agreement with those obtained in the previous studies concerning pathogen destruction by MW treatment in which complete reduction of *E. coli*, coliforms, faecal coliforms, *Ascaris lumbricoides*, etc., was achieved (Border and Rice-Spearman, 1999; Hong et al., 2004; Lamb and Siores, 2010; Mawioo et al., 2016a; Mawioo et al., 2016b).

7.8.4 Weight or volume reduction and energy consumption

The sludge weight/volume reduction results for the tested samples are shown in Figure 7-12. The weight/volume reduction increased with the increasing exposure time, and the rate of reduction varied among the sludge sample types. The highest rate of reduction was observed in the C-WAS and the lowest in the SS. However, ultimately, the highest weight/volume reduction was realized in the SS followed by WAS, C-WAS, and FS in that order, with the respective reductions of approximately 93, 90, 80, and 63%. Furthermore, the three phases that were observed and described in the temperature evolution profiles, namely the preliminary, essential, and final drying phases were also reflected in the weight reduction profiles. In all cases, the preliminary drying phase occurred within the first 30 minutes of the processing time, and was marked by a minimal weight/volume reduction. A substantial weight reduction was realized in the subsequent essential (major) drying phase, in which over 60% weight reduction was achieved. Generally this phase lasted up to approximately 90, 120, and 240 minutes for the C-WAS, FS and WAS, and SS, respectively. The final drying phase, which is depicted by a decline of the weight reduction rate was only observed in the C-WAS between 90 and 120 minutes, and WAS between 120 and 240 minutes.

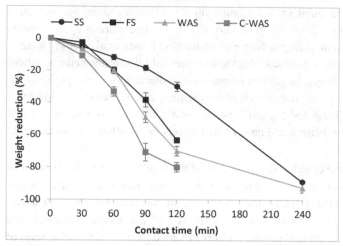

Figure 7-12. Effect of MW irradiation exposure time on the sludge weight

In order to determine the energy demand during the MW treatment of the various sludges, the weight reduction along with the input MW energy profiles were observed (Figure 7-13a). As can be seen, the trends correspond with the three drying phases that were observed in the temperature evolution and weight reduction. Furthermore, Figure 7-13b, which presents the results for the specific energy consumptions (computed by use of the actual electrical energy consumed with the corresponding weight reductions) shows relatively high but varied energy demands among the sludge samples. In each case, the highest energy demand occurred during the preliminary drying phase. For instance, (and as shown in Figure 7-13b) 18 kWh was required to achieve approximately 1 kg weight reduction in the FS, while 7-9 kWh was required to achieve approximately 1 kg reduction in the C-WAS, WAS, and SS during the preliminary phase. However, the energy consumptions at the maximum contact time were relatively lower at 3.2, 2.8, 3.5, and 4.6 kWh/kg for the C-WAS, FS, WAS, and SS, respectively. Nevertheless, they were higher than the theoretical energy demand i.e. 2.4, 2.1, 2.8, and 2.6 kWh/kg for the C-WAS, FS, WAS, and SS, respectively, that were estimated considering both the specific heat and the heat of vaporization of water.

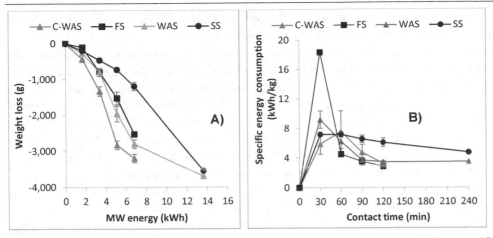

Figure 7-13. A) Sludge weight reduction and MW energy demand and, B) sludge specific energy demand as function of the MW exposure time

The order in the weight/volume reduction trends as depicted in Figure 7-12 was expected as those reductions result from vaporization of water, and are thus directly linked to the initial proportions of water and TS content in the sludge. Furthermore, the relationship between the three drying phases with the temperature profiles was anticipated since weight/volume reduction is linked to the evaporation of water, which is caused by the thermal (temperature) effects.

Generally, the results demonstrate that MW treatment was successful to achieve over 60% weight/volume reduction in all sludge types evaluated. Similar results have been achieved previously where even higher reductions over 70% were reported with the MW treatment (Menéndez et al., 2002; Mawioo et al., 2016a; Mawioo et al., 2016b). The disparity in the weight/volume reduction rates among the sludge samples was mainly attributed to their properties such as the initial water content, organic matter content, and viscosity, among others. As discussed in Section 3.2, these properties influence the temperature propagation and ultimately the moisture evaporation in the sample. Furthermore, the weight/volume reduction related drying phases (i.e. the preliminary, the essential and final drying phases) observed in this study agree with those reported in previous studies (Flaga, 2005; Mawioo et al., 2016a; Mawioo et al., 2016b). The minimum weight reduction in the preliminary phase was attributed to the high thermal capacity of water, which causes the absorption of a significant portion of the initially supplied MW energy, with a minimal temperature growth available for the evaporation of water (Mawioo et al., 2016a; Mawioo et al., 2016b). The essential drying phase, in which the majority of sludge weight/volume reduction occurs, is considered the most important stage in the MW treatment, especially if sludge weight/volume reduction is targeted. At this stage the supplied MW energy is largely used for the evaporation of the free (unbound) water that requires less energy (Mawioo et al., 2016b). As observed here with the C-WAS and FS, based on the initial thermal requirements, it is expected that the efficiency of volume reduction by MW treatment will be improved if the sludge has a higher TS content. Therefore, when possible, a pretreatment step (e.g. thickening or dewatering) can be introduced before sludge is brought to the MW treatment process. However, it would even be more prudent to promote the use of dry sanitation systems (e.g. UDDTs) in the areas that are feasibly targeted with the MW treatment (e.g. slums and emergency settlements).

Results from Figure 7-13b show that specific energy consumption is the highest during the preliminary drying phase. This can be explained by the fact that the majority of the energy supplied during this phase is largely utilized to ramp the sludge and reactor temperature with very minimal sludge weight/volume reduction. The exceptionally high energy demand registered by FS during the preliminary phase can be attributed to its stickiness (high viscosity), which is among the factors that affect the capability of the electromagnetic (MW) radiation to penetrate a material (Hong et al., 2004). The general drop of the specific energy demand in the subsequent essential drying phase can be attributed to the fact that most of the resulting heat energy was directly used for weight reduction through the vaporization of water. Ultimately, the SS exhibited the highest energy demand (i.e. 4.6 kWh/kg), which can be attributed to the relatively high water content, low amount of organic matter, and high grit (poor MW absorber) content. Conversely, the relatively low energy demand exhibited by the FS (i.e. 2.8 kWh/kg) can be linked to its low water content (implying a reduced absolute thermal capacity and amount of moisture to evaporate) and the high organic matter content which promotes irradiation. It is expected that the specific demand for FS can be reduced further if the viscosity is reduced. Moreover, the specific energy demand variations among the tested samples demonstrated the influence of the irradiated material properties on the drying rate. For instance, C-WAS and WAS, being from a similar source, demonstrated very close results regarding the specific energy consumption.

Furthermore, from the results it appears that the amount of energy input to the system during the preliminary phase is important and can influence the overall performance. For instance, if the system's total installed power is relatively low, it would take longer to increase the sludge temperature to the boiling point. Consequently, this might promote any possible heat losses to the surrounding environment. Alternatively, a relatively high total installed MW power would fasten the heating process and likely reduce the overall energy demand. Temperature ramping can also be enhanced by blending sludge with a MW receptor material, e.g. char (Menéndez et al., 2002). This is particularly viable for FS generated from dry toilet facilities (e.g. UDDTs) where the char can be applied as cover material. Application of a cover material in the dry toilet facilities is usually recommended for odor control.

The specific energy consumptions obtained here are comparable with those obtained in previous studies (Mawioo et al., 2016a; Mawioo et al., 2016b), which reported 3 kWh per kg and 2.3-2.5 kWh per kg for fresh FS and blackwater sludge, respectively, after exposure to MW irradiation. However, results from those studies and the current study are still relatively higher than those reported for the convective and conductive industrial driers which vary between 0.7–1.4 and 0.8–1.0 kWh, respectively, per kg of vaporized water (Bennamoun et al., 2013). The disparities between the results of this study and those reported in Bennamoun et al. (2013) might be only to a lesser extend attributed to different properties of materials used in the study of Bennamoun and co-workers and four sludges used in this study. It is likely due to the fact that the novel MW-based technology developed in this study was not optimized regarding the energy consumption, which will be addressed in the next stage of development. Further improvements and optimization of the unit would definitely increase the performance and ultimately reduce the energy consumption. Enhancement of the condenser (vapor extraction) unit was identified

146

as crucial to improve the efficiency of the moisture evaporation which ultimately might lower the unit's energy demand. Other factors that caused higher energy consumption in this study include the following: (i) each batch test was started with a cold reactor. A portion of the energy was initially used to warm up the reactor as well as sludge. The effect of the cold-start was not quantified in this study but can easily be checked in the future by comparison of the energy consumption between cold and hot start tests; (ii) the reactor insulation was sub-optimal. The reactor did not have sufficient thermal insulation, and this contributed to higher than desired energy consumption; and (iii) the tests were executed during the winter in cold environment (ambient temperature in the experimental hall was around 10 °C) and with cold sludge stored in the fridge. In addition, the space was very large and could not be heated up with the waste heat of the reactor. These are harsh conditions that are not representative for most of expected applications where ambient temperature is expected to be at least twice higher. These factors also contributed to the higher energy demands obtained in these study compared to the theoretical demands that were estimated considering only the specific heat and the heat of vaporization of water.

7.8.5 Energy and nutrients recovery

The MW treatment process converted the test sludge samples into three-phase products, namely solid (dried sludge), liquid (condensate), and gas. The results for the gross CV, TN, and TP measurements of the dried sludge are shown in Table 7-3. The dried FS exhibited the highest gross CV of 23 MJ/kg. C-WAS and WAS yielded 22 and 19 MJ/kg, respectively, while SS had the lowest CV of 16 MJ/kg. In addition, all the dried sludges contained expectedly a relatively high nutrient content, in which TN was generally higher than the TP across all the samples. C-WAS and WAS solids exhibited the highest TN concentration at 84 and 80 mg N/g TS, respectively; while FS and SS solids exhibited the lowest TN contents of 42 and 28 mg N/g TS, respectively. Similar trends were observed for the TP content.

Table 7-3. Characteristics of the MW irradiated sludge at maximum exposure time applied

Parameter	Sludge type			
	C-WAS	FS	WAS	SS
Water content (%)	5	8	3	2
Total solids (%)	95	92	97	98
VS/TS ratio	0.73	0.88	0.75	0.54
TN (mg/g TS)	84	42	80	28
TP (mg/g TS)	26	15	27	24
CV (MJ/kg)*	22	23	19	16
Specific energy consumption (MJ/kg)**	11.6	10.2	12.5	16.7

* Calorific value i.e. the energy contained in the dried sludge.

** The amount of energy (MJ) consumed to achieve a unit loss in sludge weight (kg).

Furthermore, the TN and TP content were measured in the condensate and the results are presented Table 7-4. In a similar trend as the solids, the condensate demonstrated high concentration of TN but a relatively low concentration of the TP.

Table 7-4. Nutrients content of the condensate from MW treated sludge

Parameter	Sludge type			
	C-WAS	FS	WAS	SS
TN (mg/L)	2,762	1,377	49	213
TP (mg/L)	0.8	0.4	0.2	0.4
COD (mg/L)	385	76	15	34

The energy values can be associated with the VS content, which represents the combustible matter in the sludge. This is why FS, having the highest organic content, had the highest CV while SS had the lowest. In addition, the CV value for WAS obtained in this study (i.e. 19 MJ/kg) is comparable to the 18.75 MJ/kg that was reported in a previous study by Chen et al. (2014). The CVs obtained here were compared with those of the common biofuels. Both coffee husks and firewood have a gross CV of 16 MJ/kg (Diener et al., 2014) which was equivalent to the SS, the lowest among the evaluated sludge samples. The CVs for the sawdust and charcoal are 20 MJ/kg and 28 MJ/kg, respectively (Diener et al., 2014), and are comparable to the FS and C-WAS.

The high nutrient contents in the C-WAS and WAS is related to the activated sludge bacteria that sequesters those nutrients from the wastewater. The SS had a higher content of TP than the FS due to the fact that septic tanks receive mixed excreta and greywater that contains phosphorus from detergents while FS was sourced from a UDDT toilet excluding the urine, which normally contains the largest fraction of phosphorus released from the body (Rose et al., 2015). Moreover, the high TN content in the condensate derived from all sludges (as indicated in Table 7-4) can be attributed to the volatilization of ammonia (NH_3) which is emitted from the hydrolysis of protein during sludge heating. Heating causes dissolution of the protein in sludge, then it hydrolyzes to form multipeptide, dipeptide and amino acid. The amino acid further hydrolyzes to form organic acid, NH_3 and CO_2 (Ren et al., 2006; Deng et al., 2009). Since NH_3 is highly soluble in water, it is expected that the volatized fraction was completely dissolved during the condensation process. The nutrients content, especially the TN in both the solids and the condensate was relatively higher than in the bio-based fertilizers such as the compost manure. For instance, some studies have reported approximately 13 mg N/g TS in the compost manure (Agegnehu et al., 2016; Andreev et al., 2016).

Generally, the energy and nutrient results demonstrated a value addition of the end products (e.g. solids and condensate) from a MW based treatment process of the various sludges. The CVs of the dried sludges were relatively higher compared to the specific energy consumptions during the treatment in a rather energy non-optimized prototype. This suggests that a substantial amount of energy can still be recovered if the sludge is used as biofuel. However, the combustion byproducts of dried sludge and their eventual treatment need to be further assessed. An alternative way for resource recovery (e.g. nutrients) from the dried sludge is by applying it as a fertilizer or soil conditioner where applicable and allowed. Similarly, the condensate has high nutrients content and can be applied as fertilizer.

7.8.6 Organic matter reduction

TS and VS fractions of the sludges were measured in the raw and treated samples and then used to compute the VS/TS values that are presented in Figure 7-14. The VS/TS ratio was used as an index to determine the organic stability of the treated sludge (Bresters et al., 1997). The results show that there was no significant change in the VS/TS ratio between the raw and the MW treated sludge samples. The final VS/TS values were 76, 88, 73%, and 55% for the C-WAS, FS, WAS, and SS, respectively.

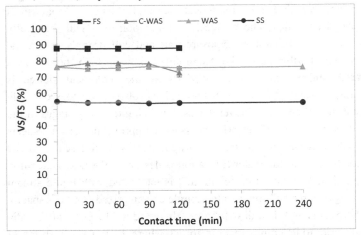

Figure 7-14. Effect of MW irradiation on the organic stability of sludge

Based on the results, it can be deduced that the MW treatment in this case expectedly did not result in organic matter reduction. The final VS/TS, especially for the C-WAS, FS, and WAS were higher than the 0.60 recommended by European Environment Agency (Bresters et al., 1997) for the sludge organic stability. However, the VS/TS ratio for the untreated SS sample was lower than 0.60; hence, the sludge was already organically stable (which is expected for SS). The results obtained here agree with those reported in the previous studies in which different kinds of sludge were heated by MW. For instance, in a study by Mawioo et al. (2016a) final VS/TS of 0.86 and 0.92 were achieved for two sample fractions of MW treated FS. In yet another study, final VS/TS values of 0.80 and 0.89 were attained with the MW treatment of two sample fractions of blackwater FS, which were in the same range as the raw sludge (Mawioo et al., 2016b). The poor organic stabilization of sludge can be attributed to the comparatively low maximum temperatures attained by the MW treatment (i.e. 102 °C) than the 550 °C normally applied for VS ignition in the gravimetric method (SM-2540E) (APHA, 2012).

Although organic matter stabilization is not achieved, the sludge was made hygienically safe for further handling and the volume was significantly reduced. As demonstrated in Section 3.5, the dried sludge samples have relatively high (carbon) energy content and can be used as biofuel to promote resource recovery. Furthermore, volume reduction can minimize handling costs during disposal or further treatment (with less costly options like composting, etc.). The organic stability may not be a major consideration in situations of intensive toilet use, e.g. slums and emergency situations, provided that pathogens are inactivated and the public health risk is reduced.

7.8.7 Future outlook of MW application in sludge treatment

The results obtained in this study clearly demonstrate the suitability of the MW-based treatment of the various kinds of sludge from the perspective of both sanitization and volume reduction.

149

The technology, when further developed and optimized particularly in terms of energy use, will be suitable for onsite applications and at the onset of the first phase of an emergency, when there are currently limited options available with the capacity to rapidly treat the large amounts of fresh sludge generated (Brdjanovic et al., 2015). In those and similar situations, pathogen inactivation should be prioritized over sludge volume reduction, as it is the most crucial aspect to curb possible excreta-related endemics. The choice for sanitization only will drastically shorten (by approximately 50 - 70 %) the sludge exposure time and will strongly reduce the energy demand of the system. Furthermore, the technology showed high potential to complement waste activated sludge treatment at WWTPs as the biological sludge is comparatively less challenging to treat than fresh faecal sludge. It can also be considered as sole technology of (pre) treatment of septic sludge. Nevertheless, this being the early stages of the research and development of the prototype, it is clearly necessary to improve the various aspects of the technology, which will increase its efficiency and applicability at full scale. At the moment, the team at IHE Delft Institute for Water Education is designing the next generation of a compact and energy efficient MW-based prototype that is integrated with liquid stream treatment and recovery of energy from dried sludge into a single containerized plant of capacity of 100 kg fresh faeces per hour. This mobile unit will be shortly tested in Slovenia and Croatia with several different sludges prior to its employment in Jordan where it will treat fresh faecal sludge from refugees in one of the camps.

7.9 Conclusions

A pilot scale MW-based prototype was developed, tested and evaluated on sanitization and volume reduction performances using four different kinds of sludge, namely partially dewatered/centrifuged waste activated sludge, fresh faecal sludge, septic tank sludge, and waste activated sludge. This study demonstrated that MW-based technology can be applied for rapid treatment of the different kinds of sludge, thereby providing a viable option for the treatment of sludge from intensively used toilet facilities (e.g. in slums and emergency settlements) and as complementary or sole treatment of waste activated sludge and/or septic tank sludge at WWTPs. A reduction below the detection limit was achieved of the pathogenic indicators. The technology resulted in a volume reduction of the raw material to over 60% and a high level of dry matter in the dried sludge was achieved (up to 98%). Besides, the process generated valuable end-products (condensate/product water and dry sludge) which can be recovered and reused as fuel, soil conditioner and/or fertilizer, etc.

Acknowledgments

This research is funded by the Bill & Melinda Gates Foundation under the framework of SaniUP project (Stimulating Local Innovation on Sanitation for the Urban Poor in Sub-Saharan Africa and South-East Asia) (OPP1029019). The authors would like to thank Tehnobiro d.o.o., Slovenia, Fricke und Mallah Microwave Technology GmbH, Germany, Ekorim, Slovenia, Maribor University, Slovenia, and the Pollution Research Group of the University of KwaZulu - Natal for their valuable support during this study.

References

Agegnehu, G., Nelson, P.N., Bird, M.I., 2016. The effects of biochar, compost and their mixture and nitrogen fertilizer on yield and nitrogen use efficiency of barley grown on a Nitisol in the highlands of Ethiopia. Sci. Total Environ. 569–570, 869-879.

Andreev, N., Ronteltap, M., Lens, P.N.L., Boincean, B., Bulat, L., Zubcov, E., 2016. Lacto-fermented mix of faeces and bio-waste supplemented by biochar improves the growth and yield of corn (Zea mays L.). Agriculture, Ecosystems & Environment 232, 263-272.

APHA, 2012. Standard Methods for the Examination of Water and Wastewater, 22nd ed. American Public Health Association, Washington DC, USA.

Banik, S., Bandyopadhyay, S., Ganguly, S., 2003. Bioeffects of microwave—a brief review. Bioresour. Technol. 87, 155-159.

Bennamoun, L., Arlabosse, P., Léonard, A., 2013. Review on fundamental aspect of application of drying process to wastewater sludge. Renew. Sust. Energ. Rev. 28, 29-43.

Bennamoun, L., Chen, Z., Afzal, M.T., 2016. Microwave drying of wastewater sludge: Experimental and modeling study. Drying Technol. 34, 235-243.

Bennamoun, L., Chen, Z., Salema, A.A., Afzal, M.T., 2014. Moisture diffusivity during microwave drying of wastewater sewage sludge. Transactions of ASABE 58, 501-508.

Border, B.G., Rice-Spearman, L., 1999. Microwaves in the laboratory: Effective decontamination. Clin. Lab. Sci. 12, 156-160.

Brdjanovic, D., Zakaria, F., Mawioo, P.M., Garcia, H.A., Hooijmans, C.M., Ćurko, J., Thye, Y.P., Setiadi, T., 2015. eSOS® – emergency Sanitation Operation System. Journal of Water Sanitation and Hygiene for Development 5, 156-164.

Bresters, A.R., Coulomb, I., Matter, B., Saabye, A., Spinosa, L., Utvik, Å.Ø., 1997. Management Approaches and Experiences: Sludge Treatment and Disposal., Environmental Issues Series 7. European Environment Agency, Copenhagen, Denmark, p. 54.

Byamukama, D., Kansiime, F., Mach, R.L., Farnleitner, A.H., 2000. Determination of Escherichia coli Contamination with Chromocult Coliform Agar Showed a High Level of Discrimination Efficiency for Differing Fecal Pollution Levels in Tropical Waters of Kampala, Uganda. Appl. Environ. Microbiol. 66, 864-868.

Chen, Z., Afzal, M.T., Salema, A.A., 2014. Microwave Drying of Wastewater Sewage Sludge. Journal of Clean Energy Technologies 2, 282-286.

Deng, W.-Y., Yan, J.-H., Li, X.-D., Wang, F., Zhu, X.-W., Lu, S.-Y., Cen, K.-F., 2009. Emission characteristics of volatile compounds during sludges drying process. J. Hazard. Mater. 162, 186-192.

Diener, S., Semiyaga, S., Niwagaba, C.B., Muspratt, A.M., Gning, J.B., Mbéguéré, M., Ennin, J.E., Zurbrugg, C., Strande, L., 2014. A value proposition: Resource recovery from faecal sludge—Can it be the driver for improved sanitation? Resources, Conservation and Recycling 88, 32-38.

Etheridge, R.D., Pesti, G.M., Foster, E.H., 1998. A comparison of nitrogen values obtained utilizing the Kjeldahl nitrogen and Dumas combustion methodologies (Leco CNS 2000) on samples typical of an animal nutrition analytical laboratory. Animal Feed Science and Technology 73, 21-28.

Flaga, A., 2005. Sludge drying, in: Plaza, E., Levlin, E. (Eds.), Proceedings of Polish-Swedish seminars. Integration and optimisation of urban sanitation systems., Cracow, Poland. Retrieved from http://www2.lwr.kth.se/Forskningsprojekt/Polishproject/index.asp?entry=13#Report13.

Haque, K.E., 1999. Microwave energy for mineral treatment processes—a brief review. Int. J. Miner. Process. 57, 1-24.

Hong, S.M., Park, J.K., Lee, Y.O., 2004. Mechanisms of microwave irradiation involved in the destruction of fecal coliforms from biosolids. Water Res. 38, 1615-1625.

Hong, S.M., Park, J.K., Teeradej, N., Lee, Y.O., Cho, Y.K., Park, C.H., 2006. Pretreatment of Sludge with Microwaves for Pathogen Destruction and Improved Anaerobic Digestion Performance. Water Environ. Res. 78, 76-83.

Ingallinella, A.M., Sanguinetti, G., Koottatep, T., Montanger, A., Strauss, M., 2002. The challenge of faecal sludge management in urban areas - strategies, regulations and treatment options. Water Sci. Technol. 46, 285-294.

Katukiza, A.Y., Ronteltap, M., Niwagaba, C.B., Foppen, J.W.A., Kansiime, F., Lens, P.N.L., 2012. Sustainable sanitation technology options for urban slums. Biotechnol. Adv. 30, 964–978.

Kim, J., Park, C., Kim, T.-H., Lee, M., Kim, S., Kim, S.-W., Lee, J., 2003. Effects of various pretreatments for enhanced anaerobic digestion with waste activated sludge. Journal of Bioscience and Bioengineering 95, 271-275.

Lamb, A.S., Siores, E., 2010. A Review of the Role of Microwaves in the Destruction of Pathogenic Bacteria, in: Anand, S.C., Kennedy, J.F., Miraftab, M., Rajendran, S. (Eds.), Medical and Healthcare Textiles. Woodhead Publishing, pp. 23-29.

Lin, Q.H., Cheng, H., Chen, G.Y., 2012. Preparation and characterization of carbonaceous adsorbents from sewage sludge using a pilot-scale microwave heating equipment. J. Anal. Appl. Pyrolysis 93, 113-119.

Mawioo, P.M., Hooijmans, C.M., Garcia, H.A., Brdjanovic, D., 2016a. Microwave treatment of faecal sludge from intensively used toilets in the slums of Nairobi, Kenya. J. Environ. Manage. 184, Part 3, 575-584.

Mawioo, P.M., Rweyemamu, A., Garcia, H.A., Hooijmans, C.M., Brdjanovic, D., 2016b. Evaluation of a microwave based reactor for the treatment of blackwater sludge. Sci. Total Environ. 548–549, 72-81.

Menéndez, J.A., Inguanzo, M., Pis, J.J., 2002. Microwave-induced pyrolysis of sewage sludge. Water Res. 36, 3261-3264.

Metcalf, Eddy, 2003. Wastewater Engineering: Treatment and Reuse, 4th ed. McGraw-Hill Publishers New York, NY. .

Park, B., Ahn, J.H., Kim, J., Hwang, S., 2004. Use of microwave pre-treatment for enhanced anaerobiosis of secondary sludge. Water Sci. Technol. 50, 17 - 23.

Remya, N., Lin, J.-G., 2011. Current status of microwave application in wastewater treatment— A review. Chem. Eng. J. 166, 797-813.

Ren, L.-h., Nie, Y.-f., Liu, J.-g., Jin, Y.-y., Sun, L., 2006. Impact of hydrothermal process on the nutrient ingredients of restaurant garbage. Journal of Environmental Sciences 18, 1012-1019.

Ronteltap, M., Dodane, P.-H., Bassan, M., 2014. Overview of Treatment Technologies, in: Strande L, Ronteltap M, Brdjanovic D (Eds.), Faecal Sludge Management - Systems Approach Implementation and Operation. IWA Publishing, London, UK, pp. 97-120.

Rose, C., Parker, A., Jefferson, B., Cartmell, E., 2015. The Characterization of Feces and Urine: A Review of the Literature to Inform Advanced Treatment Technology. Critical Reviews in Environmental Science and Technology 45, 1827-1879.

Sangadkit, W., Rattanabumrung, O., Supanivatin, P., Thipayarat, A., 2012. Practical coliforms and Escherichia coli detection and enumeration for industrial food samples using low-cost digital microscopy. Procedia Eng. 32, 126-133.

Serrano, A., Siles, J.A., Martín, M.A., Chica, A.F., Estévez-Pastor, F.S., Toro-Baptista, E., 2016. Improvement of anaerobic digestion of sewage sludge through microwave pre-treatment. Journal of Environmental Management 177, 231-239.

Takács, I., Ekama, G.A., 2008. Final Settling, in: Henze, M., Loosdrecht, M.C.M.v., Ekama, G.A., Brdjanovic, D. (Eds.), Biological Wastewater Treatment: Principles, Modelling and Design. IWA Publishing, London, UK., pp. 309-334.

Tang, B., Feng, X., Huang, S., Bin, L., Fu, F., Yang, K., 2017. Variation in rheological characteristics and microcosmic composition of the sewage sludge after microwave irradiation. Journal of Cleaner Production 148, 537-544.

Tang, B., Yu, L., Huang, S., Luo, J., Zhuo, Y., 2010. Energy efficiency of pre-treating excess sewage sludge with microwave irradiation. Bioresour. Technol. 101, 5092-5097.

Thostenson, E.T., Chou, T.W., 1999. Microwave processing: fundamentals and applications. Compos. A: Appl. Sci. Manuf. 30, 1055-1071.

Tyagi, V.K., Lo, S.-L., 2013. Microwave irradiation: A sustainable way for sludge treatment and resource recovery. Renew. Sust. Energ. Rev. 18, 288-305.

Valero, A., Cejudo, M., García-Gimeno, R.M., 2014. Inactivation kinetics for Salmonella Enteritidis in potato omelet using microwave heating treatments. Food Control 43, 175-182.

Venkatesh, M.S., Raghavan, G.S.V., 2004. An Overview of Microwave Processing and Dielectric Properties of Agri-food Materials. Biosystems Engineering 88, 1-18.

Walker, G.C., Harmon, L.G., 1966. Thermal Resistance of Staphylococcus aureus in Milk, Whey, and Phosphate Buffer. Applied Microbiology 14, 584-590.

Yu, Q., Lei, H., Li, Z., Li, H., Chen, K., Zhang, X., Liang, R., 2010. Physical and chemical properties of waste-activated sludge after microwave treatment. Water Res. 44, 2841-2849.

Chapter 8
Conclusions and outlook

8.1 Conclusions

Remarkable efforts have been deployed at different levels to improve the global sanitation provision. But there is still quite a large number of the world's population that lacks access to basic sanitation services. As at the year 2017 approximately 2.3 billion people lacked access to basic sanitation services, while another 892 million practiced open defecation (WHO/UNICEF, 2017). Reporting on sanitation provision focuses generally on the regular development context without paying specific attention to the isolated conditions such as the densely populated emergency settings and the slum settlements, where the situation is far worse. Management of human excreta is generally regarded a difficult task. But it is more complex when it involves sanitation provision in isolated situations, particularly the difficult emergency situations. Of special interest in these situations is the aspect of managing large quantities of highly contaminated fresh faecal sludge that are continuously generated from the commonly applied onsite faecal sludge containment facilities. Constraints on time and land space, etc., which are often encountered in those isolated situations add in the complexity on the entire task of sludge management. Major drawbacks in emergency sanitation can be identified as the use of less appropriate approaches and the fact that majority of the existent sanitation solutions deployed in the disaster response are not adequate to address the challenges of difficult emergency situations such as floods and high water table, unstable soils, urban and crowded areas, etc. (Johannessen, 2011).

It is expected that the effectiveness and efficiency in emergency sanitation response can greatly be improved by adopting new approaches right from the initial assessment stages through the process of choosing emergency sanitation technology options. Furthermore, sustained research and innovations in emergency sanitation can continuously expand and update the range of available technology options with those that are more tailored and adapted to the changing disaster scenarios. This can in turn minimize the impacts of disasters, especially by preventing pandemics and fatalities that are associated with poor sanitation practices in the emergency settings.

In this PhD thesis the subject of emergency sanitation was studied with the aim to gain understanding and subsequently improve on the approaches, concepts and technologies applied in the emergency settings and similar conditions. The study places emphasis on the components of containment and treatment of faecal sludge, which may be considered critical in the emergency sanitation chain. The following approaches were taken:

- Different sanitation technology options were explored with the aim to understand their potential for application in emergency situations and other similar conditions such as slums. An overview of the past emergency cases was also conducted to determine which sanitation technology options were actually applied and understand the processes that were followed in the choice of those technologies.
- An emergency sanitation concept for the faecal sludge management was proposed and developed. The concept was aimed at increasing efficiency by introducing an alternative among the traditional approaches in emergency sanitation management while promoting the aspect of providing decent sanitation to people in need and improving the entire emergency sanitation chain. The concept is mostly centered on the operations of an emergency sanitation system that makes use of ICT for monitoring and coordination.

- An alternative sludge treatment technology based on the microwave irradiation was proposed, developed and evaluated to assess its potential application for rapid treatment of faecal sludge in the emergency situations and similar contexts. As such, a review was conducted to understand the functionality and reveal any existing evidence of sanitation related applications of the microwave technology (e.g. in wastewater and/or sludge treatment). Basing on the outcomes of this review, preliminary evaluations of sludge treatment using a conventional domestic microwave unit were conducted and further expanded with a customized pilot-scale unit in which extensive practical evaluations were carried out. Furthermore, evaluations were conducted on the end products of the microwave treated sludge to assess their potential use as biofuel, soil conditioner, fertilizer, etc.

8.1.1 Review of emergency sanitation technology options and selection processes

A review of the existent sanitation technology options and emergency sanitation practices was conducted. Major resources on sanitation technologies were screened, notable among them a compilation on the compendium of sanitation technologies in emergencies by Swiss Federal Institute of Aquatic Science and Technology (EAWAG) (Robert et al., 2019) and a classification was introduced that is adapted from the compendium.

A wide range of the regular onsite and offsite sanitation options was revealed that have potential to be applied for sludge management in the emergencies. Situations with more or less similar characteristics to emergencies such as slums (which can actually be considered slow emergencies) can also benefit from these technology options. A classification was introduced that groups the technologies in either onsite or offsite categories. The merits and demerits were also identified for each technology option based on their characteristics. The onsite options are more suited for application in initial phases of an emergency situation due their ease of deployment and installation while offsite options would be preferred at the later stages when the situation has relatively stabilized. Further classification of the onsite options was introduced based on pit and non-pit options. Under the subcategory of the non-pit options was noted the open defecation option that has been applied in the field, but which we consider not suitable in the context of safe sanitation and thus recommend that its field applications should be discouraged. We noticed that some effective technologies were shunned in emergency sanitation response (even when there were no appropriate alternatives) due to their operation and maintenance costs that were considered prohibitive. For instance, during the emergency response in Haiti following the earthquake in 2010, some humanitarian actors, citing costs, discontinued the use of portaloo or chemical toilets despite their observed effectiveness. However, it is opined that given the importance of insuring public health in such crowded conditions, such expenses are worth to incur as long as the related technology is effective to address sanitation issues in place.

An overview of the specific applications of those potential technologies in the past emergencies was conducted that revealed that only a few have been applied in the field, majority of which are based on the conventional pit latrine and/or its variations and that there was a lot of replication of the options in varied scenarios. Although further research is needed to support these observations, as they are based on the few documented emergency reports that were found with specific information on sanitation, the available evidence indicates that the observed trend of technology replication even in varied scenarios can be linked to the lack of an appropriate decision making tool for technology selection. Decisions were largely influenced by individual humanitarian actors basing on their experiences in the past emergency responses and the intuition of the planners often leading to sub-optimal service delivery. There were no elaborate technology selection processes in which strong scientific evidence was used to demonstrate how a technology fits local conditions. As a contribution to address this issue, a methodology for selection of emergency sanitation technologies based on compensatory multi-criteria analysis was developed. The methodology is an important tool for decision makers involved in selection of sanitation technologies for emergency situations. The proposed methodology screens and evaluates technological options in accordance to score factors considering the capital costs, ease of deployment, space requirements, O&M costs, start-up time, use of local materials and skill requirement. An important consideration is that the proposed methodology is flexible and gives a possibility for decision makers to calculate the total weight value in accordance to prevailing conditions which are important for the disaster condition in place. Despite the aforementioned benefits the proposed methodology still needs validation under real field applications during which further developments can be expected.

In this study we also observed that most of discussion and activities around emergency sanitation response were focused on the containment facilities that do not provide any level of treatment of the faecal sludge.

In many cases, an interesting fact was the lack of clarity as to whether and how the sludge emptied from the containment facilities was treated prior to disposal in the environment. In some cases, however, it was clear that raw sludge was directly disposed of in the environment without any treatment. For instance, in Haiti during the earthquake in Port au Prince in 2010, most of the emptied sludge was disposed of in an open dumpsite and was associated with the consequent cholera outbreak. The study did not identify any strong cases to demonstrate the application of systems thinking in the emergency sanitation response. The problems mentioned above may thus be attributed to the use of ad hoc or fragmented approach to emergency sanitation that fails to holistically consider the entire range of sanitation chain components during the planning process. Therefore, we introduced a systematic concept (i.e. the emergency Sanitation Operation System (eSOS)) that would help decision makers to understand integration of the various components and link then in the decision making. The concept would ensure that all components of the sanitation chain are holistically considered right at the onset of an emergency sanitation response process.

8.1.2 The emergency sanitation operation system (eSOS)

As discussed above, the shortfalls experienced in the current emergency sanitation practices were associated with the commonly used conventional fragmented approach that does not capture the entire sanitation chain, but rather looks at the individual components separately with emphasis on the containment facilities. Based on the technology review and deliberations in the emergency sanitation conferences and workshops, it was revealed that the sanitation services based on those conventional approaches were mostly sub-optimal and that there was need to develop better approaches to enhance effectiveness and efficiency. Therefore, an innovative emergency Sanitation Operation System (eSOS) concept was introduced in this study that uses a systems approach integrating all components of an emergency sanitation chain. The concept is mostly centered on the operations of an emergency sanitation system that makes use of ICT for monitoring and coordination. It demonstrates potential to improve emergency sanitation provision and provide decent sanitation to people in need. The proposed concept presents the main components of the sanitation chain including eSOS smart toilet, an intelligent faecal sludge collection vehicle-tracking system, and a decentralized faecal sludge treatment facility. However, several critical components that required improvement were identified and explored further, including the containment and treatment. As such, an innovative treatment technology based on microwave irradiation was successfully developed and evaluated as discussed in the following section and also presented in (Chapter 5, 6, and 7) of this thesis while the eSOS smart toilet was developed and evaluated separately in another study.

By promoting a systems approach, the eSOS concept can address the bias in developing the sanitation chain components. The concept also promotes communications and coordination of emergency operations on the strength of the ICT, which has had positive transformations in other sectors. Although initial tests have successfully been conducted, the test situations were near ideal and might not have revealed all possible challenges in the system. Particularly, for the ICT component that relies on networks availability, it is prudent to carry out more rigorous evaluations of the system in the most remote locations to test its robustness and probably make further improvements. Furthermore, there is need to develop an emergency sanitation business model in order to gain a better understanding of the cost aspects related to the proposed eSOS concept.

8.1.3 Microwave irradiation based technology for rapid treatment of sludge

As a component of the proposed eSOS concept, a sludge treatment system based on the microwave irradiation technology, was developed and tested. The microwave technology study was carried out in two stages. The first stage was carried out at a laboratory scale using a domestic microwave unit and two sludge types including FS extracted from highly concentrated raw blackwater and FS from UDDT toilets. These preliminary evaluations demonstrated the capability and applicability of the technology for sludge treatment. For instance, it was shown that the technology can rapidly and efficiently reduce the sludge volume by over 70% and decrease the concentration of bacterial pathogenic indicator *E. coli* and *Ascaris lumbricoides* eggs to below the analytical detection levels. The second stage involved, more evaluations using a pilot-scale microwave reactor unit and various sludge types (i.e. waste activated sludge, faecal sludge, and septic sludge) were carried out, which similarly demonstrated that microwave treatment was successful to achieve a complete inactivation of a larger array of bacterial organisms such as *E. coli*, coliforms, *staphylococcus aureus*, and *enterococcus faecalis*) and a sludge weight/volume reduction above 60%. Besides, the dried sludge had high energy and nutrient value $\geq$ 16 MJ/kg and TN $\geq$ 28 mg/g TS and TP $\geq$ 15 mg/g TS, respectively, while the condensate had nutrient contents TN $\geq$ 49 mg/L TS and TP $\geq$ 0.2 mg/L). This revealed the potential for use of the end products as biofuel, soil conditioner, fertilizer, etc.

Microwave technology has a more rapid and more uniform volumetric heating than thermal conventional heating. The suitability of this technology for application in emergency situations or in areas with high level of faecal sludge generation can be traced from its potential for quick deployment, high level of safety and automation, etc. The technology has a better potential for complete faecal sludge sanitization since it allows a more rapid *Ascaris* eggs inactivation than sludge drying beds, composting or co-composting.

Generally, microwave treatment demonstrated potential for drying and pathogen reduction/inactivation. However, taking into consideration the relatively high energy demand, it is recommended to use the technology in urban emergencies, where land space is limited and digging up of pits is restricted. For a more efficient use, the MW reactor shall be combined with a dewatering unit e.g. centrifugation and a membrane bioreactor for the treatment of the centrate (liquid phase). This study was basically focussed on proof of concept, which was demonstrated showing effective technology in terms of volume reduction and inactivation of pathogenic organisms. Assessment of energy efficiency and optimization was not within the scope of this study, but is recommended for further investigations.

8.2 General outlook

This study highlighted the existence of numerous regular sanitation technology options with potential for application in emergency sanitation. Determination of the technologies that were applied in the past emergencies was limited since many emergency cases were documented but with either limited or no information specific to sanitation provision. Therefore, an exhaustive investigation on the sanitation technology applications in the past emergencies might be required by gathering information directly from the humanitarian actors that offered sanitation response in the documented disaster cases. Going forward the lack of adequate documentation and lack of database regarding emergency sanitation can be addressed by developing a standardized database portal dedicated to emergency sanitation and then encouraging all emergency actors to fill out and constantly update all information regarding sanitation. Humanitarian donors and governments can also help by setting as a minimum requirement to fill out those database forms by all actors for the emergency sanitation activities that they sponsor or facilitate.

The study has also demonstrated that emergency sanitation response can be improved by introducing innovations in the related sanitation approaches, tools and technologies, etc. The proposed innovative concept of *eSOS* promotes an integrated approach where all components of the sanitation chain are planned holistically. In addition, the concept demonstrates potential to reduce bias and stimulate a balanced focus such that all components of the emergency sanitation chain are developed equitably. However, there is still need to assess the cost aspects of the proposed (concept) system, which can be achieved by developing a business model

Moreover, the study underscores an innovative microwave based technology as a possible solution for the rapid treatment of sludge in the areas where large volumes are generated such as the emergency situations and the urban slums. By this technology, the massive expenditure from sanitation facilities incurred in emptying and transporting large amounts of FS generated during the disaster events and onsite sanitation facilities in densely populated areas (such as slums) and which form a major challenge to FS management can be addressed. The current study demonstrated that microwave radiation has a significant potential to reduce such costs. Being a fast, efficient and compact technology, it can reduce considerably the sludge volume and reduce the risk of excreta-related disease outbreaks by pathogen destruction, thus addressing the challenge of public health and land space constraints of urban slum and emergency settlements and producing valuable nutrient and carbon rich materials for resource recovery. However, crucial technological factors – such as the energy requirements and process optimization are important to research in order to achieve both treatment and value recovery and to optimize the efficiency of the microwave heating process. Based on the research so far, the microwave technology has a relatively high power demand compared to other drying methods such as convection and conduction. Nevertheless, this being the early stages of the research and development of the prototype, it is necessary to improve the various aspects of the technology, which will potentially increase its efficiency and applicability at full scale. Among the aspects to be improved is the condenser unit that would possibly ensure higher energy efficiency, and introduction of a unit to recover and reuse energy from the condensation process to supplement the reactor input energy requirements, both of which would potentially decrease the overall energy demand for the unit and the moisture removal efficiency and hence the weight/volume reduction. Furthermore, introduction of a unit for treatment of the water phase (via membrane filtration, etc.) as well as the odour treatment (via granulated activated carbon i.e. GAC, etc.) would ensure reuse of the water as well as reduction of the pollution thus improving the overall treatment process. The system can also be installed with higher capacity microwave generators to accelerate the process in order to shorten the residence time and ensure a more compact reactor unit. In this case, 915MHz generators can be considered owing to their high efficiencies (> 80%) of electric power to electromagnetic energy conversion and the penetration depth which is approximately three times greater than that of the 2.45GHz generators that were used in this study. Blending with different materials such as sawdust and char products, etc. and determining the best mix ratios is also recommended as it is likely to enhance the heating process by reducing the viscosity (stickiness) and/or acting as microwave receptors. Besides the technical improvements and optimizations of the unit, there is also need to further carry out more rigorous onsite field tests using wastes such as faecal from toilets located in highly dense conditions (e.g. emergency camps, refugee camps, slums, etc.) in order to enhance its experimentation for the practical applications in those conditions. Furthermore, as the unit is highly recommended for applications in isolated conditions, such as the initial phases of the emergency situations where mains power interruptions is highly likely, it is prudent to develop containerized designs (for ease of field deployment) that allow for coupling with alternative power sources such as green energy sources (e.g. the solar power) or power generating sets, and then assess their performance under the real field conditions.

Generally, emergency sanitation is still an unexplored yet exciting field for research where emergency factors make provision of sanitation even more complex. This gives a lot of research possibilities all in order to make the emergency sanitation more understood, more efficient, more affordable and appropriate to better serve millions of people in need.

References

Johannessen, Å., 2011. Identifying gaps in emergency sanitation, Design of new kits to increase effectiveness in emergencies, 2 day Workshop, 22-23 February 2011, Stoutenburg, The Netherlands. WASTE and Oxfam GB, Retrieved from http://www.susana.org/images/documents/07-cap-dev/b-conferences/13-stoutenberg-conference-2011/stoutenberg-feb-2011-report-final.pdf.

WHO/UNICEF, 2017. Progress on Sanitation and Drinking Water 2015 Update and SDG Baseline. WHO, Geneva.

About the Author

Peter Matuku Mawioo, born in 1979 in Kitui County, Kenya, obtained his Bachelor of Science (BSc) degree from the Department of Civil and Environmental Engineering; specialization in Water and Environmental Engineering in Egerton University, Kenya in 2006. In his professional career he was first engaged as project engineer in an international NGO where he served in various countries and subsequently as a graduate engineer in a civil engineering consulting firm.

In 2008, he joined IHE Delft Institute for Water Education, Delft, the Netherlands, to pursue a Master of Science (MSc) Degree in Municipal Water and Infrastructure; specialization in Sanitary Engineering. His MSc thesis was entitled "Optimization of a Novel Submerged Membrane Bioreactor System", in which he investigated the application of Membrane Bioreactor (MBR) technology for the treatment of highly concentrated Industrial Wastewater in a food processing industry located in Wijchen, Nijmegen, the Netherlands. Peter graduated in 2010 and thereafter joined the Department of Environmental Engineering and Water Technology at IHE Delft Institute for Water Education, Delft, The Netherlands as a research fellow in a water supply engineering project. The project involved investigation of various aspects of ultra-filtration and reverse osmosis systems for seawater treatment. Upon completion of the project, Peter went back to Kenya and was appointed head of water and sanitation department in a civil engineering consulting firm, his former employer. Later in 2011, he received a scholarship under the project 'stimulating local innovation on sanitation for the urban poor in sub Saharan Africa and South East Asia' funded by the Bill and Melinda Gates Foundation to pursue a PhD in the field of Sanitary Engineering at the Delft University of Technology and the IHE Delft Institute for Water Education, Delft, The Netherlands.

In 2017, he joined the Department of Civil and Environmental Engineering at Meru University of Science and Technology (MUST), Kenya as a Lecturer in Water and Wastewater Engineering. He later joined the Department of Civil and Structural Engineering at the University of Eldoret (UoE), Kenya as a Lecturer in Public Health Engineering.

List of Publications

Brdjanovic, D., Zakaria, F., Mawioo, P.M., Garcia, H.A., Hooijmans, C.M., Ćurko, J., Thye, Y.P., Setiadi, T., 2015. eSOS® – emergency Sanitation Operation System. Journal of Water Sanitation and Hygiene for Development 5, 156-164.

Mawioo, P.M., Rweyemamu, A., Garcia, H.A., Hooijmans, C.M., Brdjanovic, D., 2016. Evaluation of a microwave based reactor for the treatment of blackwater sludge. Science of the Total Environment 548–549, 72-81.

Mawioo, P.M., Garcia, H.A., Hooijmans, C.M., Brdjanovic, D., 2016. Microwave treatment of faecal sludge from intensively used toilets in the slums of Nairobi, Kenya. Jornal of Environmental Management 184, Part 3, 575-584.

Mawioo, P.M, Garcia, H.A., Hooijmans, C.M., Velkushanova, K., Simonič, M., Mijatović, I., Brdjanovic, D., 2017. A pilot-scale microwave technology based reactor for sludge drying and sanitization. Science of the Total Environment 601–602, 1437-1448.

Mawioo, P.M., Garcia, H.A., Hooijmans, C.M., Brdjanovic, D., 2018. Faecal sludge management in emergencies: A review of technology options and future perspectives. Submitted to Journal of Water Sanitation and Hygiene for Development.

Conference Contributions

Mawioo, P.M., Igbinosa E., Garcia H., Hooijmans C.M., Brdjanovic D., 2016. Emergency sanitation: A review of potential technologies and selection criteria - *In Proceedings: 3rd IWA Development Congress and Exhibition,* Nairobi, Kenya, 14-17 October 2013.

Mawioo, P.M, Garcia, H.A., Hooijmans, C.M., Velkushanova, K., Simonič, M., Mijatović, I., Brdjanovic, D., 2017. The Shit Killer - A Microwave Based Technology For Sludge Sanitizing and Drying. *Presented in the IWA Water and Development Congress and Exhibition,* Buenos Aires, Argentina, 13-16 November 2017.

Brdjanovic D., Mawioo P.M, Kocbek E., Hooijmans CM, Garcia H., Mijatovic I., 2019. A Novel Microwave-based Technology for Faecal Sludge Treatment. *Presented in the IWA Water and Development Congress and Exhibition,* Colombo, Sri Lanka, 1-5 December 2019.

Netherlands Research School for the
Socio-Economic and Natural Sciences of the Environment

D I P L O M A

For specialised PhD training

The Netherlands Research School for the
Socio-Economic and Natural Sciences of the Environment
(SENSE) declares that

Peter Matuku Mawioo

born on 25 November 1979 in Kitui, Kenya

has successfully fulfilled all requirements of the
Educational Programme of SENSE.

Delft, 16 January 2020

The Chairman of the SENSE board   the SENSE Director of Education

Prof. dr. Martin Wassen   Dr. Ad van Dommelen

The SENSE Research School has been accredited by the Royal Netherlands Academy of Arts and Sciences (KNAW)

K O N I N K L I J K E   N E D E R L A N D S E
A K A D E M I E   V A N   W E T E N S C H A P P E N

The SENSE Research School declares that Peter Matuku Mawioo has successfully fulfilled all requirements of the Educational PhD Programme of SENSE with a work load of 37.5 EC, including the following activities:

SENSE PhD Courses

o   Environmental research in context (2013)
o   Research in context activity: 'Initiating different examples of science communication on PhD research techniques and general media in Bangkok (Thailand) and in Cahul (Moldova)' (2017)

Other PhD and Advanced MSc Courses

o   WASH in emergencies, IHE Delft (2015)

External training at other research institutes

o   Training course for ML 2 laboratories, TU Delft, The Netherlands (2013)
o   Training on Analysis, Identification, Quantification and Inactivation of Helminth (Ascaris) Eggs, Asian Institute of Technology, Thailand (2014)
o   CVA Basis - Introductory Course on Industrial Safety in EU, PlusPort B.V., The Netherlands (2014)
o   Introductory e-course on Global Commons, United Nations Institute for Training and Research(UNITAR), Geneva and the University of Notre Dame, Mendoza College of Business (2012)

Management and Didactic Skills Training

o   Supervising of four MSc student with thesis (2012-2016)
o   Board Member IHE PhD Association Board (2012-2014)

Oral Presentations

o   *Novel concepts and technologies for excreta and wastewater management in challenging emergency conditions*. Integrity of Water Systems in a Developing World, 1-3 October 2012, Delft, The Netherlands
o   *Emergency sanitation technologies: A review and decision support*. Integrity of Water Systems in a Developing World, 23-24 September 2013, Delft, The Netherlands
o   *Emergency sanitation: a review of potential technologies and selection criteria*. 3rd IWA Development Congress and Exhibition, 14-17 October 2013, Nairobi, Kenya
o   *Evaluation of a microwave based reactor for faecal sludge treatment in emergency situations*. Urban sustainability, 29-30 September 2014, Delft, The Netherlands

SENSE Coordinator PhD Education

Dr. ir. Peter Vermeulen